SOLAR
ASSESSMENT GUIDANCE
A Guide for Solar Trainee, Trainer & Assessor Examination

Karthik Karuppu, Venk Sitaraman, NVICO

INDIA • SINGAPORE • MALAYSIA

Notion Press

Old No. 38, New No. 6
McNichols Road, Chetpet
Chennai - 600 031

First Published by Notion Press 2019
Copyright © NVICO Energy Private Limited 2019
All Rights Reserved.

ISBN 978-1-64650-522-7

This book has been published with all efforts taken to make the material error-free after the consent of the author. However, the author and the publisher do not assume and hereby disclaim any liability to any party for any loss, damage, or disruption caused by errors or omissions, whether such errors or omissions result from negligence, accident, or any other cause.

While every effort has been made to avoid any mistake or omission, this publication is being sold on the condition and understanding that neither the author nor the publishers or printers would be liable in any manner to any person by reason of any mistake or omission in this publication or for any action taken or omitted to be taken or advice rendered or accepted on the basis of this work. For any defect in printing or binding the publishers will be liable only to replace the defective copy by another copy of this work then available.

Contents

Editor's Preface	*v*
Author's Preface	*vii*
Eligibility Criteria for Trainee & Traineer of various levels of Solar Certification	*ix*
Acknowledgments	*xvii*
National Qualifications in Solar Energy	*xix*
Abbreviation	*xxi*
Formula	*xxvii*

Chapter 1:	Carry Out the Site Survey for Installation of Solar PV System	1
Chapter 2:	Assess the Customer's Solar PV Requirement	13
Chapter 3:	Procure the Solar PV System Components	33
Chapter 4:	Identify and Use the Tools & Tackles used for Solar PV System Installation	47
Chapter 5:	Install the Civil/Mechanical and Electrical Components of a Solar PV System	49
Chapter 6:	Pre & Post Commissioning Inspection of the Grid Connected Rooftop SPV Power Plant	65

Chapter 7:	Test and Commission Solar PV System	83
Chapter 8:	Maintain Solar PV System	123
Chapter 9:	Solar PV Project Lifecycle	133
Chapter 10:	Determine the Financial Viability of Solar PV Power Plant	135
Chapter 11:	Maintain Personal Health & Safety at Project Site	137

Answer Sheet   *141*

Bibliography   *149*

Index   *157*

Editor's Preface

In 2018, the world's energy production rebounded 2.4% based on the historical trend. International Energy Agency of USA divided the consumption patterns of various resources as coal (30%), natural gas (24%), hydro power (7%), nuclear (4%), renewable (2%) and others. The energy consumption growth remained vigorous particularly china was the main contributor to increase global energy production. USA and European Union continued to decline in global energy production by adopting strict climate policies as well as modern technologies like demand side management that slowly exit out the usage of coal, oil & gas.

The world cumulative solar PV capacity reached almost 398 GW and generated over 460 TWh till 2017, sufficient to meet 2.1% of the world's total electricity consumption. Utility-scale projects account for just over 60% of total PV installed capacity, with the rest in distributed applications such as residential, commercial and off - grid application. Over the next five years of 2018 to 2022, solar PV is expected to lead renewable electricity capacity growth, expanding by almost 580 GW. The deployment of concentrated solar power (CSP) plants is at the stage of market introduction and expansion. In 2017, the installed capacity of CSP worldwide was 4.8 GW, compared to 398 GW of Solar PV capacity. CSP capacity is expected to double by 2022 to reach 10 GW with almost all new capacity incorporating storage.

The UN Labour Agency said that 24 million new post's will be created globally by 2030, but added that "The Right Policies to promote a Greener Economy" must also be in place for this is to happen, along with better social safety nets for workers. The worldwide growth of photovoltaic is extremely dynamic and varies strongly by countries perspective and policies. The top installers of 2017 were China, United States, Japan and India. Leading PV deployments till 2018, with were China (175 GW), the United States of America (64 GW), India (25 GW), and Japan (50 GW).

Why does it hold back the growth of solar power? The reason is that the national grids are too old and cannot augment the capacity. To augment the capacity, the infrastructure investments needed are huge and the work slow. The main drawback of soalr panel is that its capacity factor is just 15% instead of the 70 to 80% capacity of the power station. A further challenge to solar system will be chronically low prices of fossil fuels, which could push back its ability to compete. Another important factor is that the solar energy can't sustain without massive subsidies. Apart from that, the benefits are less water usage, reduced air pollution, slow climate change, reduction of carbon footprint and less dependability on fossil fuels.

Author's Preface

The Ministry of New and Renewable Energy (MNRE) is responsible for research and development, intellectual property right protection, international cooperation, promotion, and coordination in renewable energy sources such as wind power, small hydro, biogas and solar power. Their main aim is to develop and deploy new and renewable energy sources for supplementing the energy requirements of the country.

MNRE has already launched a lot of renewable energy programs like Jawaharlal Nehru National Solar Mission (JNNSM), National Biogas and Manure Management Programme (NBMMP), Solar Lantern Programme LALA, Solar Thermal Energy Demonstration Programme, Remote Village Lighting Programme, National Bio-mass Cook Stoves Initiative (NBCI) and National Offshore Wind Energy Authority etc. The new initiative programs are Green Energy Corridor, Renewable Purchase Obligations, Net Metering Policy, Repowering of Wind power projects, International Solar Alliance, Suryamitra Scheme, etc.

The Ministry has established to state nodal agencies in different states and union territories of India to promote and expand the growth of efficient energy use of renewable energy. State-wise nodal agencies are NEDCAP for Andra Pradesh, TEDA for Tamil Nadu, GEDA for Gujarat, KREDL for Karnataka, RRECL for Rajasthan, TSREDCO for Telegana, etc.

By 2022, India wants to install 175 Gigawatt (GW) of renewable power capacity which corresponds to around half of its total electricity production. This is enough to replace 175 coal-fired power plants of 1,000

MW and to reduce India's dependence on fossil fuels. Fossil fuels are the source of 92% of India's electricity and they produce greenhouse gases that hasten global warming. India has set a target of achieving 40% of its total electricity generation from non-fossil fuel sources by 2030.

A blueprint draft published by Central Electricity Authority projects that 57% of the total electricity capacity will be from renewable sources by 2027. India has an installed capacity of 350 GW including renewable and non-renewable sources till 2018. As of now, Coal, 197 GW (56.4%), large hydro, 45 GW (13%), small hydro, 4.5 GW (1.3%), wind Power, 35 GW (10.1%), Solar Power, 25 GW (7.4%), Biomass, 9.2 GW (2.6%), Nuclear, 6.8GW (1.9%), Gas, 24.9 GW (7.1%) and Diesel, 0.6 GW. Till March 2018, the installed solar capacities of various states are Karnataka (5.2 GW), Telangana (3.4 GW) Andhra Pradesh (2.5 GW), Rajasthan (2.3 GW), Tamil Nadu (1.8 GW) and Gujarat (1.6 GW).

MNRE has introduced a number of Central Financial Assistance (CFA) schemes to promote solar PV in India and is to achieve its ambitious 175 GW target by 2022. The support is aimed at providing subsidies to individuals or enterprises willing to contribute to growth. Skill Council for Green Jobs (SCGJ) is developing the national occupational standards, model curriculums, and coursework activities for the area of the renewable sector. SCGJ provides the training course to the entire trainee candidates. The entire trainees will receive a certificate after successful completion of training, which will add more value to those who work in the renewable energy area.

The report cited by The Council on Energy, Environment and Water (CEEW), International Labour Organization (ILO) and the Natural Resources Defence Council (NRDC) estimates that based on surveys of solar and wind companies, developers and manufacturers, over 300,000 workers will be employed in the solar and wind energy sectors in India to meet the 2022 target.

Eligibility Criteria for Trainee & Traineer of various levels of Solar Certification

1. Solar PV Installer (Suryamitra) - Trainee

Curriculum / Syllabus

This program is aimed at training candidates for the job of a "Solar PV Installer (Suryamitra)", in the "Green Jobs" Sector/Industry and aims at building the following key competencies amongst the learner

Program Name	Solar PV Installer (Suryamitra)
Qualification Pack Name & Reference ID. ID	SGJ/Q0101, v1.0
Pre-requisites to Training	10th pass + ITI / Diploma (Electrical, Electronics, Civil, Mechanical, Fitter, Instrumentation, Welder)
Training Outcomes	**After completing this programme, participants will be able to:** • Carry out the site survey for installation of Solar PV system • Assess the customer's Solar PV requirement • Procure the Solar PV system components • Identify and Use the Tools & tackles used for Solar PV system installation • Install the Civil/Mechanical and Electrical components of a Solar PV system • Test and Commission Solar PV system • Maintain Solar PV system • Maintain personal Health & Safety at project site

Solar PV Installer (Suryamitra) - Traineer

Sr.No.	Area	Details
1	Minimum Educational Qualifications	ITI /Diploma (Electrical, Electronics, Civil, Mechanical, Fitter, Instrumentation) or B.Tech (Civil/Mechanical / Electrical / Instrumentation / Electronics / Electrical and Electronics Engineering) or MSc Physics or The education qualification can be relaxed in case of extraordinary relevant field experience

Sr.No.	Area	Details
2	Platform Certification	Recommended that the Trainer is certified for the Job Role: "Trainer", mapped to the Qualification Pack: "SSC/1402". Minimum accepted score as per SCGJ is 70%.
3	Experience	Minimum 3 years of relevant industry experience for ITI /Diploma (Electrical, Electronics, Civil, Mechanical, Fitter, Instrumentation) or Minimum 2 years of relevant industry experience for B.Tech (Civil/Mechanical / Electrical / Instrumentation / Electronics / Electrical and Electronics Engineering)

2. Solar PV Installer (Electrical) - Trainee

Curriculum / Syllabus

This program is aimed at training candidates for the job of a "Solar PV Installer – Electrical", in the "Green Jobs" Sector/Industry and aims at building the following key competencies amongst the learner

Program Name	Solar PV Installer – Electrical
Qualification Pack Name & Reference ID. ID	SGJ/Q0102, v1.0
Pre-requisites to Training	ITI / Diploma (Electrical, Electronics)
Training Outcomes	After completing this programme, participants will be able to: • Carry out the site survey for installation of Solar PV system • Identify and Use the Tools & tackles used for Solar PV system installation • Install the Electrical components of a Solar PV system • Test and Commission Solar PV system • Maintain personal Health & Safety at project site

Solar PV Installer (Electrical) - Traineer

Sr. No.	Area	Details
1	Minimum Educational Qualifications	ITI /Diploma (Electrical, Electronics) or B.Tech (Civil / Electrical / Electronics / Electrical and Electronics Engineering) or MSc Physics or The education qualification can be relaxed in case of extraordinary relevant field experience
2	Platform Certification	Recommended that the Trainer is certified for the Job Role: "Trainer", mapped to the Qualification Pack: "SSC/1402". Minimum accepted score as per SCGJ is 70%
3	Experience	Minimum 3 years of relevant industry experience for ITI /Diploma (Electrical, Electronics) or Minimum 2 years of relevant industry experience for B.Tech (Civil/Electrical/Electronics/Electrical and Electronics Engineering)

3. Solar PV Installer (Civil) - Trainee

Curriculum / Syllabus

This program is aimed at training candidates for the job of a "Solar PV Installer – Civil", in the "Green Jobs" Sector/Industry and aims at building the following key competencies amongst the learner

Program Name	Solar PV Installer – Civil
Qualification Pack Name & Reference ID. ID	SGJ/Q0103, v1.0
Pre-requisites to Training	ITI / Diploma (Electrical, Electronics, Civil, Mechanical, Fitter, Instrumentation, Welder)
Training Outcomes	**After completing this programme, participants will be able to:** • Carry out the site survey for installation of Solar PV system • Install the Civil/Mechanical components of a Solar PV system • Maintain personal Health & Safety at project site

Solar PV Installer Civil - Traineer

Sr. No.	Area	Details
1	Minimum Educational Qualifications	ITI /Diploma (Electrical, Electronics, Civil, Mechanical, Fitter, Instrumentation, welder, mason) or B.Tech (Civil/Mechanical /Electrical/ Instrumentation / Electronics / Electrical and Electronics Engineering) or The education qualification can be relaxed in case of extraordinary relevant field experience
2	Platform Certification	Recommended that the Trainer is certified for the Job Role: "Trainer", mapped to the Qualification Pack: "SSC/1402". Minimum accepted score as per SCGJ is 70%
3	Experience	Minimum 3 years of relevant industry experience for ITI /Diploma (Electrical, Electronics, Civil, Mechanical, Fitter, and Instrumentation) or Minimum 2 years of relevant industry experience for B.Tech (Civil/Mechanical / Electrical/ Instrumentation / Electronics / Electrical and Electronics Engineering)

4. Rooftop Solar Photovoltaic Entrepreneur - Trainee

Curriculum / Syllabus

This program is aimed at training candidates for the job of a "Rooftop Solar Photovoltaic Entrepreneur", in the "Green Jobs" Sector/Industry and aims at building the following key competencies amongst the learner

Program Name	Rooftop Solar Photovoltaic Entrepreneur
Qualification Pack Name & Reference ID. ID	SGJ/Q0104, v1.0
Pre-requisites to Training	B.E. / B. Tech. / Any Graduate with Science background, preferred

Training Outcomes	After completing this programme, participants will be able to: • Carry out market research and prepare a cost estimate for a Rooftop Solar Photovoltaic plant • Prepare site feasibility report • Manage Solar PV project lifecycle • Entrepreneurship skills • Maintain Personal Health & Safety at project site

Rooftop Solar Photovoltaic Entrepreneur – Traineer

Sr. No.	Area	Details
1	Minimum Educational Qualifications	B.E. / B.Tech / MSc Physics or B.Tech + MBA or B.Tech + M.Tech
2	Platform Certification	Recommended that the Trainer is certified for the Job Role: "Trainer", mapped to the Qualification Pack: "MEP/Q0102" or equivalent. Minimum accepted score as per SCGJ is 80%
3	Experience	Minimum 5 years of relevant industry experience for B.E./B.Tech / MSc Physics graduates or Minimum 3 years of relevant industry experience for (B.Tech. + M.Tech.) or (B.Tech + MBA) graduates

5. Solar Proposal Evaluation Specialist – Trainee

Curriculum / Syllabus

This program is aimed at training candidates for the job of a "Solar Proposal Evaluation Specialist", in the "Green Jobs" Sector/Industry and aims at building the following key competencies amongst the learner

Program Name	Solar Proposal Evaluation Specialist
Qualification Pack Name & Reference ID. ID	SGJ/Q0105, v1.0

Program Name	Solar Proposal Evaluation Specialist
Pre-requisites to Training	B.E. / B.Tech. / BBA / B.Com. / B.Sc. / C.A. Minimum 2 year of experience in a financial institution / bank / managing project finance for B.E. / B.Tech. / BBA / B.Com. / B.Sc. No experience required for MBA / C.A.
Training Outcomes	After completing this programme, participants will be able to: • Check the site feasibility of Solar PV Power Plant • Assess the technology feasibility of Solar PV Power Plant • Determine the financial viability of Solar PV Power Plant

Solar Proposal Evaluation Specialist – Traineer

Sr. No.	Area	Details
1	Minimum Educational Qualifications	B.E. / B.Tech. / BBA / B.Com. / B.Sc. / C.A
2	Platform Certification	Recommended that the Trainer is certified for the Job Role: "Trainer", mapped to the Qualification Pack: "MEP/Q0102" or equivalent. Minimum accepted score as per SCGJ is 80%
3	Experience	Minimum 2 projects or 20 MW of consulting or project finance experience on ground mount solar PV power plants or Minimum 10 projects or 1000 kWp of consulting or project finance experience on Rooftop solar PV power plants

6. Rooftop Solar Grid Engineer – Trainee

Curriculum / Syllabus

This program is aimed at training candidates for the job of a "Rooftop Solar Grid Engineer", in the "Green Jobs" Sector/Industry and aims at building the following key competencies amongst the learner

Program Name	Rooftop Solar Grid Engineer
Qualification Pack Name & Reference ID. ID	SGJ/Q0106, v1.0
Pre-requisites to Training	Diploma (Electrical, EEE)
Training Outcomes	After completing this programme, participants will be able to: • Pre-Commissioning Inspection of the Grid Connected Rooftop Solar PV Power Plant • Post Commissioning Testing of the Grid Connected Rooftop Solar PV Power Plant • Maintain Personal Health & safety at project site

Rooftop Solar Grid Engineer – Traineer

Sr. No.	Area	Details
1	Minimum Educational Qualifications	Diploma (Electrical, EEE) or B.Tech/B.E. (Civil, Electrical, Mechanical, Energy) or M.Tech. (Electrical, EEE, Renewable Energy)
2	Platform Certification	Recommended that the Trainer is certified for the Job Role: "Trainer", mapped to the Qualification Pack: "MEP/Q0102" or equivalent. Minimum accepted score as per SCGJ is 70%
3	Experience	Minimum 3 years of relevant industry experience for M.Tech. graduates or Minimum 5 years of relevant industry experience for B.E./B.Tech graduates or Minimum 6 years of relevant industry experience for Diploma graduates

Acknowledgments

First and foremost, I would like to thank God Almighty for giving me the strength, knowledge, ability and opportunity to undergo Solar Assessment Guidance book and to persevere and complete it satisfactorily. Without his blessings, this achievement would not have been possible.

In my journey towards this book, I have found a co-author, a friend, an inspiration, a role model and a pillar of support in publishing this book, Venk Sitaraman, Founder & CEO, NVICO Energy Private Limited. He has been there providing his heartfelt support and guidance at all times and has given me invaluable guidance, inspiration and suggestions in my quest for knowledge in cover design and publishing. He has given me all his full efforts in proof reading the book, while silently and non-obtrusively ensuring that I stay on core subject and do not deviate from the core of the Solar Assessment Guidance.

My acknowledgement would be incomplete without thanking the biggest source of my strength, my family. The blessings of my parents Mr. K. Muniyandi & Mrs. M. Seethai and the love and made a tremendous contribution in helping me reach this stage in my life. I thank them for putting up with me in difficult moments where I felt stumped and for goading me on to follow my dream of getting this book completion. This would not have been possible without their unwavering and unselfish love and support given to me at all times.

Karthik Karuppu

National Qualifications in Solar Energy

SCGJ has developed the following Qualifications:

Sr. No.	Name of the QP	Code	NSQF Level
1	Solar PV Installer (Suryamitra)	SGJ/Q0101	4
2	Solar PV Installer (Electrical)	SGJ/Q0102	4
3	Solar PV Installer (Civil)	SGJ/Q0103	4
4	Rooftop Solar Photovoltaic Entrepreneur	SGJ/Q0104	6
5	Solar Proposal Evaluation Specialist	SGJ/Q0105	7
6	Rooftop Solar Grid Engineer	SGJ/Q0106	5

Abbreviation

5S	–	Each term starts with an S
%	–	Percentage
η	–	Efficiency

A

AC	–	Alternating Current
ACB	–	Air Circuit Breaker
AH	–	Amphere Hour

B

BOOT	–	Build, Own, Operate, Transfer
BIS	–	Bureau of Indian Standards
BoS	–	Balance of System
BOM	–	Bill of Materials

C

CFA	–	Central Financial Assistance
CAPEX	–	Capital Expenditure
CdTe	–	Cadmium Telluride
CIGS	–	Copper Indium Gallium Selenide
CC	–	Charge Controller
CFL	–	Compact Fluorescent Lamp

°C	–	degree Celsius
CEIG	–	Chief Electrical Inspector to Government
C Rating	–	Capacity Rating
CB	–	Combiner Box

D

DOD	–	Depth of Discharge
DC	–	Direct Current
DT	–	Distribution Transformer
DAS	–	Data Acquisition Systems

E

EB	–	Electricity Bill
ESCO	–	Energy Service Company

F

FF	–	Fill Factor

G

GaInP	–	Gallium Indium Phosphide
GI	–	Galvanized Iron

H

Hr	–	Hour
HT/LT	–	High / Low Tension
HV Test	–	High Voltage Test

I

IP	–	Ingress Protection
Im	–	Current at Maximum Power Point
Isc	–	Short Circuit Current
ISO	–	International Organization for Standardization

IEC	–	International Electrotechnical Commission
I/A	–	Current / Amphere
IR Test	–	Insulation Test
Irms	–	Root Mean Square Current
IV Curve	–	Current Voltage Curve
IRR	–	Internal Rate of Return
IREDA	–	Indian Renewable Energy Development Agency

J

JB	–	Junction Box
JNNSM	–	Jawaharlal Nehru National Solar Mission

K

Kmph	–	Kilometer per hour
KW	–	Kilo Watt
kΩ	–	Kilo Ohm
KVA	–	Kilo Volt Amphere
KV	–	Kilo Volt
KWh	–	Kilo Watt Hour

L

LA	–	Lighting Arrestor
LCA	–	Life Cycle Assessment
LCC	–	Life Cycle Cost

M

MEP	–	Mechanical, Electrical, Plumbing
MPPT	–	Maximum Power Point Tracking
MS Excel	–	Microsoft Excel
MS Word	–	Microsoft word
MS PPT	–	Microsoft Power Point Presentation
MC4	–	Multi Contact and 4 for the 4mm Diameter Contact Pin

MNRE	–	Ministry of New and Renewable Energy
MW	–	Mega Watt
MCB	–	Miniature Circuit Breaker
MWh	–	Mega Watt Hour

N

NAPCC	–	National Action Plan on Climate Change
NSDC	–	National Skill Development Corporation
NSQF	–	National Skills Qualifications Framework
NOCT	–	Normal Operating Cell Temperature

O

OPEX	–	Operating Expenditure
O&M Cost	–	Operation and Maintenance Cost
OHS	–	Occupational Health and Safety
O &M	–	Operation & Maintenance
OCB	–	Oil Circuit Breaker

P

PW	–	Pega Watt
PV Plant	–	Photo Voltaic Plant
PV Module	–	Photo Voltaic Module
PPA	–	Power Purchase Agreement
Pmax	–	Maximum Power Point
PR	–	Performance Ratio
P	–	Power
PPE	–	Personal Protective Equipment
Pm	–	Maximum Power

Q

QA	–	Quality Assurance
QA	–	Quality Control

R

RESCO	–	Renewable Energy Service Company
RMS	–	Root Mean Square
R	–	Resistance
ROI	–	Return on Investment

S

SMF	–	Sealed Maintenance Free
SSC	–	Sector Skill Council
Sqft	–	Square Foot
Sqm	–	Square Meter
SECI	–	Solar Energy Corporation of India
SCGJ / SGJ	–	Skill Council for Green Jobs
STC	–	Standard Temperature Condition
SOP	–	Standard Operating Procedure
SPD	–	Surge Protection Device
SF6	–	Sulphur Hexafluoride
SoC	–	State of Charge

T

TW	–	Tera Watt
TEDA	–	Tamilnadu Energy Development Agency

U

UL	–	Underwriters Laboratories
UPS	–	Uninterrupted Power Supply
UV	–	Ultra Violet

V

Vrms	–	Root Mean Square Voltage
V-P Characteristics	–	Voltage Power Characteristics

V_{oc}	–	Open Circuit Voltage
V_m	–	Voltage at Maximum Power Point
V	–	Voltage
V_{drop}	–	Voltage drop

W

Wh	–	Watt hour
Wp	–	Watt Peak
W/m^2	–	Watt per Square Meter

Formula

- Voltage drop (Vd) = Resitance (R) × Current (I)
- Power (W) = Voltage (V) × Current (I)
- Electrical energy (Whr) = Power (W) × Time (hr)
- DC power (Pdc) = Current (Idc) × Voltage (Vdc)
- Root mean square current (Irms) = Maximum peak current (Im) / $\sqrt{2}$
- Apparent power (Papp) = Volatge (Vrms) × Current (Irms)
- Actual / Real power (Preal) = Apparent Power (Papp) × Power Factor (PF)
- Fill factor (FF) = Im Vm / Isc Voc
- Fill factor (FF) = Pm / Isc Voc
- Efficiency = Maximum power (Pmax) / Input Power (Pin) x Area
- Voc (New reduced Voltage) = Voc (STC) − (Vdec X (T1-T2))
- η (New reduced efficiency) = η (STC) − (ηdec X (T1-T2))
- P (New reduced Power) = P (STC) − (Pdec X (T1-T2))
- Battery Capacity (Ah) = Current (A) × Hour (h)
- Energy storage in battery = Voltage (V) × Capacity (Ah)
- DoD (%) = 100% − SoC (%)
- SoC (%) = 100% − DoD (%)
- C-rating = Capacity / No. of hours for full charge or discharge
- Inverter efficiency = Output / Input

- Inverter output power = Efficiency x Input Power
- Power loss of inverter = Loss = Input – Output
- Inverter input DC power = V × I
- Efficiency of inverter = V × I x PF / V × I
- Resistance R = ρ (L / A)
- Battery Efficiency = Output energy / Input energy
- DC voltage of series string = DC voltage of individual solar PV module × number of modules connected in series
- DC current of series connected solar PV string = DC current of individual solar PV module
- DC power output of series connected solar PV string = DC voltage of series connected solar PV string × DC current of series connected solar PV string
- AC power feed to grid = DC Power produced from PV array – Power Lost in converting DC to AC power
- DC current of parallel connected solar PV string = DC current of individual solar PV module × Number of modules connected in parallel
- DC voltage of parallel connected solar PV string = DC voltage of individual solar PV module
- DC power output of parallel-connected solar PV string = DC voltage of parallel-connected solar PV string × DC current of parallel-connected solar PV string
- DC voltage of series-parallel connected solar PV array = DC voltage of series connected string
- DC current of series-parallel connected solar PV array = DC current of individual PV module string x number of strings to be connected in parallel
- DC power output of series-parallel connected solar PV array = DC voltage of series-parallel connected solar PV array × DC current of series-parallel connected solar PV array

- PV array DC power output = Inverter DC power input / (1 − DC cable losses)
- Inverter DC power input = Inverter AC power output / (1 − Inverter power losses)
- AC power output of inverter = Power fed to load / (1 − AC power losses)
- Performance ratio = Electricity fed to grid / Electricity generated by plant in absence of any losses
- Inverter output AC power = $V \times I \times PF$
- Total power $P = P1 + P2 + P3$, if Module no. 1, 2, 3 are P1, P2, P3 in Watts
- Total $I_{sc} = I_{sc1} + I_{sc2}$, if two modules are in parallel combination
- Total battery voltage $V = V1 + V2$, if Two battery voltage of V1 & V2 are connected in series
- Total battery voltage $V = V1 + V2 + V3$, if Three battery voltages of V1, V2 & V3 are connected in series
- Total current $I = I1 = I2 = I3$, if Three battery current of I1, I2 & I3 are connected in series
- Total battery voltage $V = V1 = V2$, if Two battery voltages of V1 & V2 are connected in Parallel
- Total current $I = I1 + I2 + I3$, if Three battery current of I1, I2 & I3 are connected in parallel

Chapter 1

Carry Out the Site Survey for Installation of Solar PV System

1. Solar constant value is _____
   a) 1.366 kW / m²
   b) 2.366 kW / m²
   c) 3.366 kW / m²
   d) 4.366 kW / m²

2. In India, solar panel should be faced _____ direction due to northern hemisphere
   a) Northward
   b) Southward
   c) Westward
   d) Eastward

3. In Australia, solar panel should faced _____ direction due to southern hemisphere
   a) Northward
   b) Southward
   c) Eastward
   d) Westward

4. _____ face of PV module to give the optimum output
   a) North
   b) South
   c) East
   d) West

5. Global solar radiation is the sum of _____
   a) Direct
   b) Diffuse
   c) Reflected
   d) All of these

6. Direct solar radiation passes directly through _____
   a) The atmosphere to the Earth's surface
   b) Only atmosphere
   c) Only earth surface
   d) None

7. Diffuse solar radiation is _____ in the atmosphere
   a) Scattered
   b) Reflected
   c) Diffused
   d) None

8. Reflected solar radiation reaches a surface and is reflected to _____
   a) Adjacent surfaces
   b) Atmosphere
   c) Earth
   d) None

9. Solar radiation at night _____
   a) Zero
   b) Minimum
   c) Maximum
   d) None

10. _____ has effect on PV plant construction
    a) Soil Quality
    b) Wind Speed
    c) Option a & b
    d) None

11. _____ is depend on tilt angle
    a) Latitude
    b) Azimuth
    c) Sea level
    d) Roof pitch

12. Site survey including of _____
    a) Access of roof
    b) Water supply connection
    c) EB substation
    d) All of these

13. Shadow is due to _____
    a) Building
    b) Tree
    c) Option a & b
    d) None

14. Fixed structure of PV module can result in _____ power generation than single axis mechanical tracker
    a) Less
    b) More
    c) Option a & b
    d) None

15. The difference of expected and actual output of the PV system depends on _____
    a) Solar Irradiation
    b) Tilt Angle
    c) Option a & b
    d) None

16. _____ factor depends on site feasibility study
    a) Soil Test
    b) Wind Speed
    c) Solar Irradiance
    d) All of these

17. _____ factor depends on site feasibility study
    a) Roof Pitch
    b) Location
    c) Substation
    d) all of these

18. _____ factor depends on site feasibility study
    a) Shadow analysis
    b) Total Land Area
    c) Water Quality
    d) All of these

19. _____ is not a part of Site feasibility study
    a) Soil Quality
    b) Population of nearest city
    c) Wind Speed
    d) Location

20. _____ factor is considered to design a residential roof top solution
    a) Connected Load
    b) Total available area
    c) Tariff Structure
    d) All of these

21. _____ factor is not considered to design a residential roof top solution
    a) Total number of person
    b) Nearest load center
    c) Collection of electricity bill
    d) Battery back up hours

22. _____ is the following factor needed in designing PV plant
    a) Location
    b) Longitude and Latitude
    c) Type of roof
    d) All of these

23. _____ is the following factor needed in designing PV plant
    a) Shadow Analysis   b) Available roof area
    c) Tilt Angle   d) All of these

24. _____ is the following factor needed in designing PV plant
    a) Nearest Substation   b) Site Layout
    c) Soil Report   d) All of these

25. _____ is the following factor needed in designing PV plant
    a) Roof Pitch   b) EB Bill
    c) Tariff type   d) All of these

26. _____ is the following factor which affects the solar PV output
    a) Shading Obstacles   b) Direction of Sun
    c) Roof Pitch of the PV system d) All of these

27. What will happen for the PV module output power while using single / double axis mechanical tracker than fixed structure?
    a) Increases   b) Decreases
    c) Option a & b   d) None

28. The PV module output power _____ if latitude of location is not matched with tilt angle
    a) Increases   b) Decreases
    c) Option a & b   d) None

29. _____ type of mechanical tracker is used for PV mounting structure
    a) Single axis tracker   b) Double axis tracker
    c) Option a & b   d) None

30. _____ factor affects the solar PV output
    a) Sun Direction   b) Shadow of building
    c) Location of PV panel placed d) All of these

31. _____ is the one which will optimize the PV plant design
    a) Price of components
    b) Site location
    c) Quality of the PV system
    d) All of these

32. _____ is the affecting factor of PV output
    a) Weather Condition
    b) Quality of PV system
    c) Option a & b
    d) None

33. _____ refers the solar module tracker
    a) PV Mounting System
    b) Inverter Mounting System
    c) Battery Mounting System
    d) None

34. Current generated from the shaded cell will be _____ than the rest of the solar cells
    a) Lower
    b) Higher
    c) Same
    d) None

35. Shaded cell will always _____ in the current flow
    a) Resist
    b) Non Resist
    c) Option a & b
    d) None

36. Non shaded cell always generate _____
    a) Full Current
    b) Half Current
    c) No Current
    d) None

37. Shaded cell act as a _____
    a) Load
    b) Current
    c) Voltage
    d) None

38. Shaded solar cell can become _____
    a) Very hot
    b) Too Cool
    c) Moderate
    d) None

39. Hot spot result in _____
    a) Shaded cell
    b) Non shaded cell
    c) Option a & b
    d) None

40. Solar current generation depends on _____
    a) Light Falls
    b) Heat Energy
    c) Option a & b
    d) None

41. Solar voltage generation independent of the _____
    a) Cell Area
    b) Amount of light falls
    c) Angle of light falls
    d) All of these

42. Maximum open circuit voltage depends on _____
    a) Operating Temperature
    b) Light Intensity
    c) Angle of Light Falls
    d) None

43. Electricity generated by solar cell depends upon the _____
    a) Intensity of Light
    b) Area of a cell
    c) Angle at which light falls on it
    d) All of these

44. Current generation of cell depends on the _____
    a) Cell area
    b) Amount of solar radiation falling on cell
    c) Angle of cell
    d) All of these

45. _____ is the factor affecting the power generation of solar cell
    a) Amount of light
    b) Cell Area
    c) Day light falling angle
    d) All of these

46. Amount of electricity generated is directly proportional to the amount of _____
    a) Light falling
    b) Temperature
    c) Heat falling
    d) None

47. _____ is the factor which affects the power generation of solar cell
   a) Operating Temperature   b) Soil Deposit
   c) Conversion Efficiency   d) All of these

48. More power generation is due to _____
   a) Higher light falls   b) Temperature is equal to STC at 25 °C
   c) Option a & b   d) None

49. Lower power generation is due to _____
   a) Lower light falls   b) Temperature is more than STC value
   c) Option a & b   d) None

50. PV output is _____ if not considered tilt angle
   a) Lower   b) Higher
   c) Remain same   d) None

51. Power generation from PV system is affected by _____
   a) Climate Variation   b) Tilt Angle
   c) Option a & b   d) None

52. When sunlight falls perpendicular to the surface of a module, it always gives _____
   a) Maximum Power   b) Minimum Power
   c) Option a & b   d) None

53. When the light does not fall perpendicular on the module, it always gives _____
   a) Less Output Power   b) More Output Power
   c) Option a & b   d) None

54. PV module output is reduced by _____
    a) Module Temperature Loss
    b) Module Soiling Loss
    c) Option a & b
    d) None

55. The amount of power generation throughout the day keeps _____
    a) Changing
    b) Constant
    c) Not Changing
    d) None

56. Normally in India, the daily solar radiation varies between _____
    a) 4 to 6 kWh /m² /day
    b) 8 to 10 kWh /m² /day
    c) 9 to 12 kWh /m² /day
    d) 12 to 20 kWh /m² /day

57. _____ area is required for 1 KW solar plant
    a) 100 sq. ft
    b) 150 sq. ft
    c) 200 sq. ft
    d) 250 sq. ft

58. _____ is the area required for a 1 KW solar system
    a) 10 Sq. m
    b) 15 Sq. m
    c) 20 Sq. m
    d) 25 Sq. m

59. _____ is the area required for a 1 MW solar system
    a) 0.5 to 1 Acre
    b) 1 to 3 Acre
    c) 4 to 6 Acre
    d) 8 to 12 Acre

60. _____ unit is generated for 1 KW system per year
    a) 1600 to 1700
    b) 2000 to 3000
    c) 3000 to 4000
    d) 4000 to 5000

61. _____ is the approximate cost for 1 MW solar plant
    a) 4 to 6 Crore
    b) 8 to 10 Crore
    c) 11 to 12 Crore
    d) None

62. In one day _____ unit of electricity is generated from 1 KW solar plant
    a) 4-6
    b) 8-10
    c) 12-14
    d) 16-18

63. In one day _____ unit of electricity is generated from 1 MW solar plant
    a) 4000-6000
    b) 8000-10000
    c) 12000-14000
    d) 16000-18000

64. _____ months have more solar radiation
    a) January to February
    b) April to May
    c) October to November
    d) None

65. _____ months have low solar radiation
    a) March to April
    b) April to May
    c) November to December
    d) None

66. _____ months have average solar radiation
    a) April to May
    b) July to August
    c) November to December
    d) None

67. At 1000 W/m² solar irradiance, if peak output power is 230 Wp then at 400 W/m² solar irradiance, the peak output power will be _____

    Formula: New maximum output power = (Available maximum output power value / actual solar irradiance value) x Design solar irradiance value under STC
    Note : (230/1000) x 400
    a) 29
    b) 92
    c) 115
    d) 105

68. At 1000 W/m² solar irradiance, if peak output power is 230 Wp then at 200 W/m² solar irradiance, the peak output power will be _____

   Formula : New maximum output power = (Available maximum output power value / actual solar irradiance value) x Design solar irradiance value under STC
   Note : (230/1000) x 200
   a) 26 b) 36
   c) 46 d) 64

69. At 1000 W/m² solar irradiance, if peak output power is 230 Wp then at 500 W/m² solar irradiance, the peak output power will be _____

   Formula : New maximum output power = (Available maximum output power value / actual solar irradiance value) x Design solar irradiance value under STC
   Note : (230/1000) x 500
   a) 105 b) 115
   c) 195 d) 511

70. At 800 W/m² solar irradiance, if peak output power is 230 Wp then at 1000 W/m² solar irradiance, the peak output power will be _____

   Formula : New maximum output power = (Available maximum output power value / actual solar irradiance value) x Design solar irradiance value under STC
   Note : (230/800) x 1000
   a) 280 b) 282
   c) 288 d) 828

71. At 400 W/m² solar irradiance, if peak output power is 230 Wp then at 1000 W/m² solar irradiance, the peak output power will be _____
    Formula : New maximum output power = (Available maximum output power value / actual solar irradiance value) x Design solar irradiance value under STC
    Note : (230/400) x 1000
    a) 575 b) 725
    c) 735 d) 755

72. A solar PV module's maximum power output at 200 W/m² and 400 W/m² is 100 watt and 200 watt respectively. What will be the PV Wp rating of the module under STC ?
    Formula : New maximum output power = (Available maximum output power value / actual solar irradiance value) x Design solar irradiance value under STC
    Note: (100/200) x 1000), (200/400) x 1000
    a) 250, 250 b) 500, 500
    c) 750, 750 d) 1000, 1000

73. _____ is a site survey details
    a) Roof Design b) Building Orientation
    c) Available roof space d) All of these

Chapter 2

Assess the Customer's Solar PV Requirement

1. JNNSM means _____
   a) Jawaharlal Nehru National Solar Mission
   b) Jawaharlal Nehru National Solar Mirror
   c) Jawaharlal Nehru National Solar Measurements
   d) Jawaharlal Nehru National Solar Message

2. NAPCC means _____
   a) National Action Plan on Climate Change
   b) National Action Plan on Climate Condition
   c) National Action Plan on Climate Chance
   d) National Action Plan on Climate Combat

3. MNRE means _____
   a) Ministry of New and Renewable Energy
   b) Ministry of Non Renewable Energy
   c) Option a & b
   d) None

4. SECI means _____
   a) Solar Energy Corporation of India
   b) Solar Electricity Corporation of India
   c) Solar Economy Corporation of India
   d) Solar Education Corporation of India

5. IREDA means _____
   a) Indian Renewable Energy Development Agency
   b) International Renewable Energy Development Agency
   c) Iceland Renewable Energy Development Agency
   d) Ireland Renewable Energy Development Agency

6. The state nodal agency of solar project in Tamilnadu is _____
   a) Tamilnadu Energy Development Agency
   b) Telangana Energy Development Agency
   c) Option a & b
   d) None

7. NSDC means _____
   a) National Skill Development Corporation
   b) National Skill Development Cooperation
   c) National Skill Development Curriculum
   d) None

8. SCGJ means _____
   a) Skill Council for Green Jobs
   b) Skill Corporation for Green Jobs
   c) Skill Company for Green Jobs
   d) Skill Count for Green Jobs

9. NSQF means _____
   a) National Skills Qualifications Framework
   b) National Skills Qualified Framework
   c) National Skills Qualifications Frame
   d) National Skills Qualifications Family

10. The SCGJ qualification pack is _____
    a) Solar PV Installer - Suryamitra
    b) Solar PV Installer - Electrical
    c) Solar PV Installer - Civil
    d) All of these

11. The SCGJ qualification pack _____
    a) Rooftop Solar Grid Engineer
    b) Solar Proposal Evaluation Specialist
    c) Rooftop Solar Photovoltaic Entrepreneur
    d) All of these

12. _____ code is used for solar PV Installer (Suryamitra) qualification pack
    a) SGJ/Q0101
    b) SGJ/Q0102
    c) SGJ/Q0103
    d) SGJ/Q0104

13. The NSQF level _____ is used for solar PV Installer (Suryamitra) qualification pack
    a) 4
    b) 5
    c) 6
    d) 7

14. _____ qualification pack is referred for SGJ/Q0101
    a) Solar PV Installer - Suryamitra
    b) Solar PV Installer - Electrical
    c) Solar PV Installer - Civil
    d) None

15. The code _____ is used for solar PV Installer (Electrical) qualification pack
    a) SGJ/Q0102
    b) SGJ/Q0101
    c) SGJ/Q0103
    d) SGJ/Q0104

16. The NSQF level _____ is used for solar PV Installer (Electrical) qualification pack
    a) 4
    b) 5
    c) 6
    d) 7

17. _____ qualification pack is referred for SGJ/Q0102
    a) Solar PV Installer - Electrical
    b) Solar PV Installer - Suryamitra
    c) Solar PV Installer - Civil
    d) None

18. The code _____ used for solar PV Installer (Civil) qualification pack
    a) SGJ/Q0103
    b) SGJ/Q0101
    c) SGJ/Q0102
    d) SGJ/Q0104

19. The NSQF level _____ is used for solar PV Installer (Civil) qualification pack
    a) 4
    b) 5
    c) 6
    d) 7

20. _____ qualification pack is referred for SGJ/Q0103
    a) Solar PV Installer - Civil
    b) Solar PV Installer - Electrical
    c) Solar PV Installer - Suryamitra
    d) None

21. The code _____ is used for rooftop solar photovoltaic entrepreneur qualification pack
    a) SGJ/Q0104
    b) SGJ/Q0101
    c) SGJ/Q0103
    d) SGJ/Q0102

22. The NSQF level _____ is used for rooftop solar photovoltaic entrepreneur qualification pack
    a) 1
    b) 2
    c) 3
    d) 6

23. _____ qualification pack is referred for SGJ/Q0104
    a) Rooftop Solar Photovoltaic Entrepreneur
    b) Rooftop Solar Grid Engineer
    c) Solar Proposal Evaluation Specialist
    d) None

24. The code _____ is used for Rooftop Solar Grid Engineer qualification pack
    a) SGJ/Q0101
    b) SGJ/Q0102
    c) SGJ/Q0103
    d) SGJ/Q0106

25. The NSQF level _____ is used for solar PV Installer (Suryamitra) qualification pack
    a) 4 b) 5
    c) 6 d) 7

26. The NSQF level _____ is used for Rooftop Solar Grid Engineer qualification pack
    a) 1 b) 4
    c) 5 d) 6

27. _____ qualification pack is referred for SGJ/Q0106
    a) Rooftop Solar Grid Engineer
    b) Rooftop Solar Photovoltaic Entrepreneur
    c) Solar Proposal Evaluation Specialist
    d) None

28. The code _____ is used for Solar Proposal Evaluation Specialist qualification pack
    a) SGJ/Q0101 b) SGJ/Q0102
    c) SGJ/Q0103 d) SGJ/Q0105

29. The NSQF level _____ is used for Solar Proposal Evaluation Specialist qualification pack
    a) 4 b) 5
    c) 6 d) 7

30. _____ qualification pack is referred for SGJ/Q0105
    a) Solar Proposal Evaluation Specialist
    b) Rooftop Solar Photovoltaic Entrepreneur
    c) Rooftop Solar Grid Engineer
    d) None

31. A trainer must have the certification of "SSC/1402" is eligible to take training to the trainee for the following qualification pack
    a) Solar PV Installer - Suryamitra
    b) Solar PV Installer - Electrical
    c) Solar PV Installer - Civil
    d) All of these

32. SSC/ 1402 qualification pack is applicable to _____
    a) Trainer   b) Trainee
    c) Worker   d) Labour

33. A trainer must have the certification of "MEP/Q0102" is eligible to take training to the trainee for the following qualification pack _____
    a) Rooftop Solar Grid Engineer
    b) Solar Proposal Evaluation Specialist
    c) Rooftop Solar Photovoltaic Entrepreneur
    d) All of these

34. MEP/Q0102 qualification pack is applicable to _____
    a) Trainer   b) Trainee
    c) Worker   d) Labour

35. CAPEX means _____
    a) Capital Expenditure   b) Captive Extension
    c) Cash Exchange   d) None

36. OPEX means _____
    a) Operating Expenditure   b) Operating Equipment
    c) Open Exchange   d) None

37. CAPEX is known as _____
    a) Expenditure incurred in assets
    b) Expenditure needed to run its business
    c) Option a & b
    d) None

38. OPEX is known as _____
    a) Expenditure needed to run its business
    b) Expenditure incurred in assets
    c) Option a & b
    d) None

39. BOOT means _____
    a) Build, Own, Operate, Transfer
    b) Build, Operate, Own, Translate
    c) Build, Obey, Own, Translate
    d) Build, Option, Oppose, Temporary

40. RESCO means _____
    a) Renewable Energy Service Company
    b) Renewed Energy Service Company
    c) Renewable Electricity Service Company
    d) Renewable Energy Service Corporation

41. RESCO is an ESCO Energy service company which _____
    a) Provides energy to the consumers from Renewable Energy Sources
    b) Provides energy to the consumers from Non Renewable Energy Sources
    c) Option a & b
    d) None

42. CFA means _____
    a) Central Financial Assistance
    b) Centre Financial Audit
    c) Certified Financial Accreditation
    d) None

43. In India 2022, rooftop sector target national solar mission for _____
    a) 5 GW
    b) 12 GW
    c) 22 GW
    d) 40 GW

44. Solar installation sets target of _____ by 2022 in India
    a) 80 GW
    b) 100 GW
    c) 140 GW
    d) 150 GW

45. _____ is the total installed power generation capacity in India till 2018
    a) 350 GW
    b) 750 GW
    c) 850 GW
    d) 1000 GW

46. The world to be achieved 100% renewable sources positively by _____
    a) 2025
    b) 2100
    c) 2080
    d) 2050

47. The country which utilized the most wind energy in 2018 is _____
    a) Russia
    b) France
    c) Sri Lanka
    d) Japan

48. _____ is the share of renewable energy in Indian power sector in terms of installed capacity
    a) 15.6 %
    b) 25.6 %
    c) 29.9 %
    d) 36.5 %

49. _____ is the share of renewable energy in Indian power sector in terms of energy generation
    a) 6.6 %   b) 9.6 %
    c) 16.1 %   d) 21.3 %

50. _____ is the share of fossil energy in Indian power sector in terms of energy generation
    a) 28.7 %   b) 81.9 %
    c) 98.4 %   d) 100 %

51. The country which is using 100 % renewable energy is _____
    a) Iceland   b) New Zealand
    c) Ireland   d) Finland

52. Renewable sources of iceland are_____
    a) Hydro, Geo thermal   b) Wind, Solar
    c) Biogas, Thermal   d) None

53. The country which utilized the most solar energy in 2018 is _____
    a) China   b) Japan
    c) USA   d) Germany

54. _____ is the world leader in the number of solar power systems installed per capita
    a) Germany   b) Japan
    c) USA   d) Canada

55. Advantage of the solar system is _____
    a) Unlimited Source   b) Environmental friendly
    c) Option a & b   d) None

56. _____ is the advantage of solar system
    a) No air pollution
    b) No water pollution
    c) No greenhouse gases
    d) All of these

57. _____ is the disadvantage of solar system
    a) Initial cost is high
    b) Weather dependent
    c) Uses a lot of space
    d) All of these

58. The disadvantage of solar system is _____
    a) Energy storage is expensive
    b) O&M cost is high
    c) Life cycle cost is high
    d) All of these

59. Disadvantage of the solar system is _____
    a) Interrupted power supply
    b) High initial cost
    c) Tariff rate is high
    d) All of these

60. _____ is correct for V-I Characteristics of PV module
    a) Voltage ∝ 1 / Current
    b) Voltage ∝ Power x Current
    c) Voltage ∝ Power / Current
    d) None

61. _____ is correct for V-P Characteristics of PV module
    a) Voltage ∝ Power
    b) Voltage ∝ Power x Current
    c) Voltage ∝ Current x Power
    d) Voltage ∝ 1 / Power

62. Voltage _____, if PV modules are connected in series
    a) Increases
    b) Decreases
    c) option a & b
    d) None

63. Current _____, if PV modules are connected in parallel
    a) Increases
    b) Decreases
    c) Option a & b
    d) None

64. Power _____, if PV modules are connected either in series / parallel
    a) Remains same
    b) Increases
    c) Decreases
    d) None

65. A DC circuit is a circuit in which current flows in only _____
    a) One direction
    b) Two Direction
    c) All direction
    d) No direction

66. In AC circuit, current flows in _____ the direction
    a) Both
    b) One
    c) Option a & b
    d) None

67. The direction of current does not change with time is called _____
    a) AC
    b) DC
    c) Both a & b
    d) None

68. The variation of current with respect to time is called _____
    a) AC
    b) DC
    c) Both a & b
    d) None

69. PV module generates _____
    a) AC
    b) DC
    c) AC & DC
    d) None

70. The categories of PV system is _____
    a) Standalone Solar PV systems
    b) Grid - connected Solar PV system
    c) Hybrid Solar PV system
    d) All of these

71. _____ component does not have in grid - connected solar PV system
   a) Battery
   b) Inverter
   c) PV module
   d) MPPT

72. _____ component does have in Standalone solar PV systems
   a) Battery
   b) Net Meter
   c) Option a & b
   d) None

73. Standalone Solar PV system is called as _____
   a) OFF Grid Solar PV systems
   b) Grid - connected Solar PV system
   c) Hybrid Solar PV system
   d) None

74. Grid - connected solar PV system is called as _____
   a) ON Grid Solar PV systems
   b) OFF Grid Solar PV system
   c) Hybrid Solar PV system
   d) None

75. Hybrid solar PV system is the combination of _____
   a) ON & OFF Grid Solar PV systems
   b) ON Grid Solar PV systems
   c) OFF Grid Solar PV system
   d) None

76. Single solar cell can generate daily electricity in range of _____
   a) 6 Wh to 10 Wh
   b) 10 Wh to 15 Wh
   c) 20 Wh to 30 Wh
   d) 40 Wh to 50 Wh

77. The energy density of solar radiation is approximately _____
   a) 1000 W/m²
   b) 2000 W/m²
   c) 2000 W/m²
   d) 3000 W/m²

78. The average peak sun hours may actually be closer to _____
    a) 3 or 4
    b) 7 to 10
    c) 10 to 12
    d) None

79. _____ are the loads to be considered in solar PV system
    a) AC & DC loads
    b) Only AC loads
    c) Only DC loads
    d) None

80. The solar cell materials is made up of _____
    a) Mono - Crystalline Silicon
    b) Amorphous Silicon
    c) Cadmium Telluride
    d) All of these

81. CdTe means _____
    a) Cadmium Telluride
    b) Calcium Telluride
    c) Cadmium Theoretical
    d) Calcium Theoretical

82. CIGS means _____
    a) Copper Indium Gallium Selenide
    b) Copper Indium Gas Selenide
    c) Cadmium Indium Gallium Selenide
    d) Copper Indium Gas Silicon

83. GaInP means _____
    a) Gallium Indium Phosphide
    b) Gas Indium Phosphide
    c) Gallium Indium Polycrystalline
    d) Galvanized Indium Phosphide

84. Thin film solar cell efficiency is _____
    a) 6 to 9 %
    b) 14 to 17 %
    c) 18 to 20 %
    d) 20 to 22 %

85. Efficiency of CdTe and CIGS solar cell is _____
    a) 8 to 11 %   b) 12 to 15 %
    c) 16 to 18 %   d) 20 to 22 %

86. Poly or multi - crystalline solar cell efficiency is _____
    a) 13 to 16 %   b) 70 to 79 %
    c) 89 to 92 %   d) 94 to 98 %

87. Mono - crystalline silicon solar cell efficiency is _____
    a) 15 to 20 %   b) 86 to 89 %
    c) 89 to 92 %   d) 93 to 95 %

88. Efficiency of Gallium Indium Phosphide or Gallium Arsenide or Germanium solar cell is _____
    a) 30 to 35 %   b) 50 to 59 %
    c) 60 to 70 %   d) 80 to 90 %

89. Crystalline Silicon Fill Factor range is _____
    a) 30 to 40 %   b) 40 to 50 %
    c) 70 to 80 %   d) 90 to 95 %

90. _____ material has higher efficiency
    a) Mono Crystalline   b) Thin Film
    c) Polly Crystalline   d) None

91. Solar panels can still produce _____ of their typical output on a cloudy day
    a) 10 to 25 %   b) 30 to 40 %
    c) 40 to 55 %   d) None

92. Battery delivers in _____ form
    a) AC   b) DC
    c) Square   d) None

93. _____ is the best method to store solar energy
   a) Battery
   b) Inverter
   c) Transformer
   d) None

94. The input power of inverter is _____
   a) AC
   b) DC
   c) Square
   d) None

95. The output power of inverter is _____
   a) AC
   b) DC
   c) Square
   d) None

96. The maximum efficiency of inverter is _____
   a) 30 to 40 %
   b) 50 to 60 %
   c) 70 to 75 %
   d) 90 to 98 %

97. _____ type of mounting system gives a maximum power output
   a) Solar trackers
   b) Fixed mounting
   c) Seasonal tilt for every four month
   d) None

98. MPPT is helpful to deliver _____ to the systems
   a) Maximum Power
   b) Minimum Power
   c) Option a & b
   d) None

99. PV module will not deliver _____ without MPPT
   a) Maximum Power
   b) Minimum Power
   c) Option a & b
   d) None

100. Site hand over should be done with _____
   a) In the presence of electrical inspector
   b) Proper Documentation & Check Lists
   c) Both a & b
   d) None

101. _____ factor should be checked before installing the PV plant
    a) Easy access of roof
    b) Easy access of transport
    c) Water availability
    d) All of these

102. _____ is the way to educate a customer about solar system
    a) Free energy from sun
    b) No environmental pollution
    c) Little maintenance
    d) All of these

103. _____ is the way to educate a customer about solar system
    a) Infinite sources of energy
    b) Fixed tariff till 25 years
    c) Independent of grid supply
    d) All of these

104. Customer should be aware of _____
    a) Abnormal Condition
    b) Operating Procedure
    c) Data Logging
    d) All of the above

105. Electrical connection of PV plant is approved by _____
    a) CEIG
    b) Contractor
    c) Customer
    d) None

106. _____ meter is preferred for Roof Top installation
    a) Net
    b) Gross
    c) Option a & b
    d) None

107. To whom should I get a permission to do civil work on the roof for installing the PV plant is _____
    a) Contractor
    b) Customer
    c) Electricity Board
    d) All of these

108. Who is the main authority for giving certificate for solar PV installer ?
    a) NSDC / SCGJ
    b) BHEL
    c) NTPC
    d) TNEB

109. _____ provides net meter
    a) Electricity Board
    b) Collector Office
    c) Gram Panchayat
    d) Public Works Department

110. Who is giving subsidies for solar Plant ?
    a) MNRE/State Nodal Agencies
    b) NLC
    c) TANGEDCO
    d) TANTRANSCO

111. Gross meter measures _____
    a) All generated energy from PV plant is directly injected to grid without allowing any self-consumption
    b) All generated energy from PV plant is directly injected to grid with allowing of self-consumption
    c) Option a & b
    d) None

112. Solar net meter is measures the _____
    a) Excess energy export to grid
    b) Deficit energy import from grid
    c) Option a & b
    d) None

113. Net meter measures the _____
    a) Self consumption plus excess solar generation is injected to grid
    b) Only self consumption
    c) Only excess solar generation injected to grid
    d) None

114. Net meter measures _____
    a) Both total imported energy and total exported energy
    b) Only imported energy
    c) Only exported energy
    d) None

115. Who is state nodal agencies in Tamilnadu ?
    a) TEDA
    b) NREDCAP
    c) KREDL
    d) ANERT

116. _____ is the limited factor under net metering regulation
    a) Distribution Transformer Load
    b) Connected load
    c) Electricity Tariff
    d) All of these

117. PPA means _____
    a) Power Purchase Agreement
    b) Purchase Power Agreement
    c) Power Plant Agreement
    d) Power Phase Agreement

118. A 2 BHK house is installed the rooftop capacity of 10 kWp after assessing the site visit . Take the average solar radiation value of 4 kWh/m²/day. Assume 10 % losses due to soil deposit, temperature, snow fall, dust, shadow, voltage drop, cable loss etc., Find the total units delivered to the customer house ?
    a) 25 units
    b) 34 units
    c) 36 units
    d) 40 units

119. A 0.3 acre of industrial owner is installed rooftop capacity of 50 kWp after site survey. Take the average solar radiation value of 5 kWh/m²/day. Assume 10 % for generation, 5 % for inverter & other 5 % transmission losses. Find the total units delivered to the industrial owner ?
    a) 200 units
    b) 250 units
    c) 340 units
    d) 400 units

120. Microsoft Excel is useful for _____
    a) Calculation
    b) Charts
    c) Option a & b
    d) None

121. Microsoft Word is useful for writing _____
 a) Letters
 b) Essay
 c) Option a & b
 d) None

122. Microsoft Power Point is useful for creating a _____
 a) Graphics
 b) Presentation
 c) Movies
 d) All of these

Chapter 3

Procure the Solar PV System Components

1. _____ device is used for converting sunlight into electrical energy
   a) Solar Panel
   b) Solar Water Heater
   c) Iron Box
   d) None

2. _____ device is used for converting solar radiation energy into heat energy
   a) Solar Collector
   b) Solar Water Heater
   c) Iron Box
   d) None

3. Pyranometer is used to measure the average solar radiation range from _____ kWh /m² /day
   a) 4 to 6
   b) 8 to 10
   c) 11 to 22
   d) 12 to 14

4. _____ number of solar panels are needed for 1 MW pre commissioned plant if it designed for 300 W solar module
   a) 3333
   b) 4444
   c) 5555
   d) 6666

5. Pyranometer is used to measure _____
   a) Radiation
   b) Wind speed
   c) Air mass
   d) Ambient Temperature

6. _____ is used to measure the ambient temperature
   a) Thermometer
   b) Temperature
   c) Temporary
   d) None

7. Anemometer is used to measure the _____
   a) Wind Speed
   b) Wind Direction
   c) Option a & b
   d) None

8. Temperature sensor is used to measure _____
   a) Ambient Temperature
   b) Cell Temperature
   c) Option a & b
   d) None

9. IEC 61215 is applicable for _____
   a) PV module
   b) Inverter
   c) Structure
   d) None

10. IEC 61646 is the standard for _____
    a) Thin Film Module
    b) Poly Crystalline Module
    c) Mono Crystalline Module
    d) Both b & c

11. IEC 61215 is the standard for _____
    a) Poly Crystalline Module
    b) Mono Crystalline Module
    c) Option a & b
    d) None

12. The IEC code _____ is used for Concentrator Photo Voltaic (CPV) module
    a) 62108
    b) 61215
    c) 61646
    d) 61730

13. The IEC code _____ is used for Photo Voltaic module safety qualification
    a) 62730
    b) 61215
    c) 61646
    d) 61108

14. _____ standard is applicable for flat plate Photo Voltaic module
    a) UL 1703
    b) UL 1803
    c) UL 3001
    d) None

15. _____ IEC code is used for inverter
    a) 62109
    b) 61215
    c) 61646
    d) 61108

16. A solar panel's performance warranty will typically guarantee _____ production at 10 years
    a) 90 %
    b) 80 %
    c) 70 %
    d) 100 %

17. A solar panel's performance warranty will typically guarantee _____ production at 25 years
    a) 80 %
    b) 90 %
    c) 70 %
    d) 100 %

18. Panels will degrade by about _____ each year
    a) 1 %
    b) 2 %
    c) 3 %
    d) 4 %

19. Degradation of crystalline solar modules over 10 years is _____
    a) 1 %
    b) 5 %
    c) 10 %
    d) 20 %

20. PV module warranty is for at least _____ years
    a) 5
    b) 10
    c) 15
    d) 25

21. _____ connector is used to take the PV output terminal
    a) MC4   b) LC4
    c) ML4   d) CM4

22. In solar PV system, structure should be designed to withstand wind loads corresponding wind speeds up to
    a) 150 kmph   b) 400 kmph
    c) 500 kmph   d) 520 kmph

23. Galvanization thickness of about _____ is good enough for structure
    a) 10 to 20 micron   b) 20 to 40 micron
    c) 20 to 60 micron   d) 80 to 100 micron

24. Charge Controller (CC) is to protect the life of _____
    a) Battery   b) Inverter
    c) MPPT   d) PV Module

25. MPPT means _____
    a) Maximum Power Point Tracking
    b) Maximum Power Point Transfer
    c) Minimum Point Power Tracker
    d) Maximum Power Point Tester

26. _____ is the electronic based tracker of PV module
    a) MPPT   b) LPPT
    c) CPPT   d) KPPT

27. _____ is very much helpful to extract maximum power at any condition
    a) MPPT   b) Charge Controller
    c) Inverter   d) Rectifier

28. _____ controller is required in PV systems to protect the batteries
    a) Charge b) Node
    c) Mesh d) None

29. _____ controller is required to avoid battery over charge
    a) Charge b) Node
    c) Mesh d) None

30. Size of the battery is referred as _____
    a) A b) AA
    c) AAA d) All of these

31. Size of the battery is not to be referred _____
    a) AAAA b) C
    c) D d) A

32. Battery is made up of _____
    a) Lead Acid b) Nickel Cadmium
    c) Nickel Metal Hydride d) All of these

33. _____ kind of battery is used for solar storage
    a) Lithium Ion b) Lithium Ion Polymer
    c) Option a & b d) None

34. SMF battery means _____
    a) Sealed Maintenance Free b) Sealed Maintain Free
    c) Sealed Mounted Free d) Sealed Minimum Free

35. _____ device is used to store the electrical energy
    a) Battery b) Inverter
    c) Rectifier d) Controller

36. _____ is used to store the electrical energy from solar module
    a) Battery   b) MPPT
    c) Inverter   d) UPS

37. _____ is not required for grid connected solar system
    a) Battery   b) Inverter
    c) MPPT   d) Structure

38. _____ is required for standalone solar system
    a) Battery   b) Rectifier
    c) Option a & b   d) None

39. _____ device is required to convert DC to AC
    a) Inverter   b) Rectifier
    c) Computer   d) Mobile Phone

40. The device which is used to store the electrical energy is _____
    a) Battery   b) Rectifier
    c) Inverter   d) None

41. _____ energy is stored in batteries
    a) Chemical   b) Mechanical
    c) Wave   d) Air

42. _____ is required to convert DC to AC
    a) Inverter   b) Battery
    c) Rectifier   d) UPS

43. _____ is the type of grid connected inverter
    a) Central Inverter   b) String Inverter
    c) Module / Micro Inverter   d) All of these

44. Anti-isolating protection is applicable to _____
    a) Inverter
    b) PV Module
    c) Net Meter
    d) None

45. IEC Code for anti-islanding grid connected inverter is _____
    a) 62116
    b) 66116
    c) 62548
    d) 61215

46. The function of anti-islanding protection in a grid connected inverter is to _____
    a) Disconnect from Grid
    b) Disconnect from PV Module
    c) Disconnect from PV Module
    d) None

47. Anti-isolating protection is applicable to _____
    a) Inverter
    b) PV module
    c) Net Meter
    d) None

48. DAS means _____
    a) Data Acquisition Systems
    b) Direct Assessment System
    c) Direct Affected Systems
    d) None

49. Inverter response to frequency fluctuation is _____
    a) 49 Hz to 51 Hz
    b) 47.5 Hz to 50.5 Hz
    c) 45 Hz to 55 Hz
    d) 47.5 Hz to 52.5 Hz

50. Inverter power loss is _____ in solar power plant
    a) 2 to 4 %
    b) 15 to 17 %
    c) 10 to 15 %
    d) None

51. The combiner box has a _____
    a) Fuse
    b) Blocking Diode
    c) Surge Protection
    d) All of these

52. The Combiner Box is to _____
    a) Connect from multiple strings of PV Module to Inverter
    b) Connect from string Inverter to Battery
    c) Connect from Battery to Transformer
    d) None

53. _____ is connected from all the string of PV modules to inverter
    a) Combiner Box   b) Connected Box
    c) Common Box   d) None

54. The Combiner Box is helpful to _____ during maintenance period
    a) Isolate between PV String and Inverter
    b) Isolate between Transformer and Inverter
    c) Isolate between Transformer and PV string
    d) None

55. Junction box is certified by _____
    a) ISO   b) BIS
    c) IEC   d) All of these

56. Outdoor installed inverter rated is _____
    a) IP 54   b) IP 56
    c) IP 65   d) IP 75

57. ISO means _____
    a) International Organization for Standardization
    b) Indian Organization for Standardization
    c) International Orientation for Standardization
    d) Inter Organization for Standardization

58. BIS means _____
   a) Bureau of Indian Standards
   b) Bureau of International Standards
   c) Best of Indian Standards
   d) Before of Indian Standards

59. IEC means _____
   a) International Electrotechnical Commission
   b) International Electrotechnical Commission
   c) Inter-carrier Electrotechnical Commission
   d) International Electromechanical Commission

60. AC & DC cable loss is _____ in solar power plant
   a) 2 to 5 %
   b) 18 to 20 %
   c) 10 to 15 %
   d) None

61. _____ is the component of solar system
   a) PV Module
   b) Combiner Box
   c) Inverter
   d) All of these

62. _____ is the component of solar system
   a) Structure
   b) AC/DC DB
   c) AC/DC Cable
   d) All of these

63. Components of the Solar PV system is _____
   a) AC Combiner Box
   b) Pyranometer
   c) Lighting Arrestor
   d) All of the above

64. Components of the Solar PV system is _____
   a) MC4 Connector
   b) DC Combiner Box
   c) Anemometer
   d) All of the above

65. The component of solar system is _____
    a) Switchgear   b) Lighting Arrestor
    c) Earth Pit   d) All of these

66. _____ is the component involved in PV system
    a) Inverter   b) MPPT & Charge Controller
    c) Battery   d) All of these

67. LA means _____
    a) Lighting Arrestor   b) Level Area
    c) Light Area   d) Lifting Area

68. SPD is used for _____
    a) Surge Protection   b) Under Voltage
    c) Over Current   d) Ground Fault

69. SPD means _____
    a) Surge Protection Device   b) Single Protective Device
    c) Sound Protection Device   d) Signal Protection Device

70. The equipment's protection available in PV plant is known as _____
    a) Over Current   b) Over / Under Voltage
    c) Short Circuit   d) All of these

71. The equipment's protection available in PV plant is known as _____
    a) Over / Under Frequency   b) Earth Fault
    c) Surge   d) All of these

72. The equipment's protection available in PV plant is known as _____
    a) Anti – isolating   b) Earth Fault
    c) Surge   d) All of these

73. The equipment's protection available in PV plant is known as _____
    a) Master Trip Relay   b) Anti Pumping
    c) Transformer Relay   d) All of these

74. _____ type of protection is used for Solar Photo Voltaic system
    a) Circuit Breaker   b) Surge Protection Device
    c) Lightning Arrestor   d) All of these

75. The purpose of transformer is to _____
    a) Step up the Voltage   b) Step down the Voltage
    c) Option a & b   d) None

76. _____ instrument is used to measure air speed
    a) Anemometer   b) Thermometer
    c) Pyrometer   d) None

77. _____ is known as circuit breaker
    a) MCB   b) OCB
    c) ACB   d) All of these

78. SF6 means _____
    a) Sulphur Hexafluoride   b) Sulphur Fluoride
    c) Option a & b   d) None

79. BoS means _____
    a) Balance of System   b) Boot of System
    c) Battery of System   d) None

80. The BoS of the PV system is _____
    a) Switch Gear   b) Earthing Kit
    c) AC and DC cables   d) All of these

81. The BoS of the PV system is _____
    a) Cable
    b) Distribution Board
    c) Lighting Arrestor
    d) All of these

82. _____ component is existed in the PV plant
    a) Battery
    b) Charge Controller
    c) MPPT
    d) All of above

83. _____ component is existed in the PV plant
    a) PV Module
    b) Structure
    c) Inverter
    d) All of above

84. The component of the solar system is _____
    a) Battery
    b) MPPT
    c) Share Controller
    d) All of these

85. In India "Earth / Ground" wire colour refers to _____
    a) Green
    b) Blue
    c) Brown
    d) None

86. The conductive material is _____
    a) Copper
    b) Aluminium
    c) Option a & b
    d) None

87. Aluminum has more _____ than copper
    a) Resistivity
    b) Capacitivity
    c) Option a & b
    d) None

88. In India "Neutral" wire colour refers _____
    a) Black
    b) Blue
    c) Brown
    d) None

89. _____ is responsible for completing the PV system
   a) Contractor
   b) Customer
   c) Government in Charge
   d) All of these

90. The net metering is purchased from _____
   a) Electricity Board
   b) Customer
   c) Contractor
   d) None

91. BOM means _____
   a) Bill Of Materials
   b) Business Operating Method
   c) Best Operating Manuals
   d) Business Operating Model

Chapter 4

Identify and Use the Tools & Tackles used for Solar PV System Installation

1. The essential assessment tool required for site is _____
   a) Measuring Tape b) Compass
   c) Digital Camera d) All of these

2. _____ is the essential tool required for installation
   a) Wire strippers b) Crimper
   c) Cutters d) All of these

3. _____ is the essential tool required for installation
   a) Chalk line b) Multimeter
   c) Hole Saw & Punch d) All of these

4. _____ is the essential tool required for installation
   a) Torque wrench b) Pliers
   c) Hack Saw d) All of these

5. _____ is the essential tool required for installation
   a) Extension Boards b) Drill Machine
   c) Fuse Puller d) All of these

6. _____ is the essential tool required for installation
   a) Set of Spanners b) Set of Screwdrivers
   c) Set of Crimping tools d) All of these

7. _____ is the essential tool which is required for battery system
   a) Hydrometer or Refractometer
   b) Small Flashlight
   c) Rubber Apron
   d) All of these

8. _____ is the essential tool which is required for battery System
   a) Rubber Gloves
   b) Safety Goggles
   c) Baking Soda
   d) All of these

9. The purpose of flash light using in battery system is _____
   a) To view electrolyte level
   b) To view corrosive on battery terminals
   c) Option a & b
   d) None

10. Baking soda is used in battery system is _____
    a) To neutralize any acid spills
    b) To eliminate the corrosive on battery terminals
    c) Option a & b
    d) None

11. The measuring tool is _____
    a) Measuring Tape
    b) Line Dori
    c) Option a & b
    d) None

12. _____ is a measuring tool
    a) Vernier Caliper
    b) Plumb Bob
    c) Option a & b
    d) None

Chapter 5

Install the Civil/Mechanical and Electrical Components of a Solar PV System

1. The flow of current is the _____ in an electrical circuit
   a) Flow of Electrons
   b) Flow of Neutrons
   c) Flow of Protons
   d) None

2. The greater amount of current flows when the _____ is higher
   a) Voltage
   b) Momentum
   c) Force
   d) None

3. The flow of charge (electrons) through a wire per second is called _____
   a) Electric Current
   b) Electrical Voltage
   c) Electrical Power
   d) Electrical Energy

4. The amount of current flows through a wire depends on the _____
   a) Voltage
   b) Diameter of the wire
   c) Both a & b
   d) None

5. Higher diameter of wire means _____ to current flow
   a) Low Resistance
   b) High Resistance
   c) Both a & b
   d) None

6. Diameter of the wire which indicates the resistance to the _____
   a) Flow of Current
   b) Flow of Neutrons
   c) Flow of Protons
   d) None

7. The amount of voltage drop depends on _____ and _____ flowing through the wire
   a) Resistance, Current
   b) Voltage, Power
   c) Resistance, Voltage
   d) Capacitance, Voltage

8. If the resistance of a wire is 0.5 Ohm and current flowing through the wire is 5 A then the Voltage drop is _____
   a) 0.025 V
   b) 0.25 V
   c) 2.5 V
   d) 25 V

9. The interconnection of the various electrical components is known as _____
   a) Electrical Circuit
   b) Electronic Circuit
   c) Internet
   d) Web

10. The unit of current is _____
    a) Ampere
    b) Ohm
    c) Watts
    d) Farad

11. The unit of Voltage is _____
    a) Volt
    b) Ohm
    c) Watts
    d) Farad

12. The unit of Power is _____
    a) Watts
    b) Ohm
    c) Henry
    d) Farad

13. Power is defined as _____
    a) Energy / Time
    b) Energy x Time
    c) Energy + Time
    d) Energy – Time

14. One Unit means _____
    a) 10 W
    b) 100 W
    c) 1000 W
    d) 10000 W

15. _____ has the maximum power
    a) 100 W
    b) 200 W
    c) 400 W
    d) 500 W

16. _____ is the smallest electrical energy
    a) 1 Wh
    b) 10 Wh
    c) 100 Wh
    d) 1000 Wh

17. The unit for Electrical energy is _____
    a) Watt Hour
    b) Watts
    c) Ampere
    d) Volt

18. One Joule is equivalent to _____
    a) 1 Watt Second
    b) 1 Watt Hour
    c) Option a & b
    d) None

19. One kWh is equivalent to _____
    a) 3006 KJ
    b) 3060 KJ
    c) 3600 KJ
    d) 3603 KJ

20. 1000 W is equivalent to _____
    a) 1 KW
    b) 10 KW
    c) 100 KW
    d) 1000 KW

21. 1 MW is equivalent to _____
    a) 1 KW
    b) 10 KW
    c) 100 KW
    d) 1000 KW

22. In AC circuit, the actual power delivered to the load is normally _____ than the apparent power
    a) Less
    b) More
    c) Equal
    d) None

23. AC means _____
    a) Alternating Current
    b) Alternative Circuit
    c) Analyzed Current
    d) None

24. DC means _____
    a) Direct Current
    b) Diffused Current
    c) Discontinued Current
    d) None

25. Car battery voltage level is _____
    a) 12 V
    b) 120 V
    c) 220 V
    d) 330 V

26. House voltage level is _____
    a) 230 V
    b) 600 V
    c) 800 V
    d) 1000 V

27. Sub station voltage level is _____
    a) 11 KV
    b) 21 KV
    c) 31 KV
    d) 41 KV

28. Torch light battery level is _____
    a) 1.5 V
    b) 15 V
    c) 150 V
    d) 1500 V

29. Renewable energy is _____
    a) Solar
    b) Wind
    c) Biomass
    d) All of these

30. Non - Renewable energy is _____
    a) Oil and Gas
    b) Petroleum
    c) Coal
    d) All of these

31. Standalone system is applicable for _____
    a) Small KW
    b) Large MW
    c) Option a & b
    d) None

32. Grid connected system is applicable for _____
    a) Large MW system
    b) Micro watts system
    c) Mini watts system
    d) None

33. PV plant has _____ maintenance than thermal power plant
    a) Less
    b) More
    c) Option a & b
    d) None

34. Feed In Tariff (FIT) is applicable for _____
    a) Renewable Energy
    b) Non Renewable Energy
    c) Option a & b
    d) None

35. The solar irradiance is measured in _____
    a) W / m^2
    b) W^2 / m
    c) m^2 / W
    d) $W \times m^2$

36. Unit of solar PV power is rated as _____
    a) Watt Peak
    b) Watt
    c) Watt Hour
    d) None

37. Watt Peak is related to _____
    a) PV Plant
    b) Thermal Plant
    c) Hydro Plant
    d) None

38. Peak solar irradiation value is always _____ than average solar irradiance
    a) More
    b) Less
    c) Option a & b
    d) None

39. Average solar irradiation value is always _____ than peak solar irradiance
    a) Less
    b) More
    c) Option a & b
    d) None

40. The average solar irradiance value is _____, if its peak solar radiance is 6 kWh / sq.mt / day
    a) 5
    b) 7
    c) 8
    d) 9

41. The peak solar irradiance value is _____, if its average solar radiance is 4 kWh / sq.mt / day
    a) 5
    b) 3
    c) 1
    d) 2

42. Photovoltaic panel is directly converts _____ into electricity
    a) Sunlight
    b) Heat
    c) Option a & b
    d) None

43. Smaller cell area produces _____
    a) Small current
    b) Large Current
    c) Option a & b
    d) None

44. Large area solar cell produces _____
    a) Large current
    b) Small Current
    c) Option a & b
    d) None

45. Mono crystalline cell voltage range is _____
    a) 0.5 to 0.65 V
    b) 1 to 2 V
    c) 3 to 4 V
    d) 5 to 7 V

46. Area of the solar cell has increased while the current is _____
    a) Increased
    b) Same
    c) Decreased
    d) None

47. High temperature in solar cell results in _____
    a) Lower Power Output
    b) Higher Power Output
    c) Same Power Output
    d) None

48. NOCT means _____
    a) Normal Operating Cell Temperature
    b) Nominal Operating Cell Temperature
    c) Normal Own Cell Temperature
    d) Normal Output Cell Temperature

49. Typical PV module NOCT is specified at _____
    a) 15 °C
    b) 25 °C
    c) 45 °C
    d) 55 °C

50. If cell temperature in PV module increases, the voltage output also _____
    a) Decreases
    b) Increases
    c) Remains same
    d) None

51. If cell temperature in PV module increases, the module efficiency also _____

   a) Decreases   b) Increases
   c) Remains same   d) Non

52. If cell temperature in PV module increase, the output power also _____

   a) Decreases   b) Increases
   c) Remains same   d) None

53. The short circuit current of PV module _____ with increase in cell temperature

   a) Increases   b) Decreases
   c) Remains same   d) None

54. The value of Vm if Voc of 30 V is _____
    Note : Normally Vm is about 80 % to 85 % of Voc

   a) 20 to 22   b) 24 to 25
   c) 30 to 32   d) 40 to 42

55. The value of Im if Isc of 8 A is _____
    Note : Normally Im is about 90 % to 95 % of Isc

   a) 7.2 to 7.6   b) 8 to 8.2
   c) 9 to 9.2   d) 10 to 10.2

56. The maximum current (Im) will always be _____ than short circuit current (Isc)

   a) Lower   b) Higher
   c) Remains Same   d) None

57. The maximum current (Im) is equal to about _____ the short circuit current (Isc) of the module
   a) 90 to 95 %
   b) 76 to 80 %
   c) 70 to 75 %
   d) 60 to 65 %

58. The maximum voltage (Vm) will always be _____ than open circuit voltage (Voc)
   a) Lower
   b) Higher
   c) Remain Same
   d) None

59. The maximum voltage (Vm) is equal to about _____ of the open circuit voltage (Voc) of the PV module
   a) 80 to 85 %
   b) 66 to 70 %
   c) 50 to 55 %
   d) 60 to 65 %

60. The value of Pin = 1000 W/ m^2, Im = 0.5 A, Vm = 0.8 V, Area = 0.0026 m^2. The efficiency is _____
   a) 15 %
   b) 18 %
   c) 21 %
   d) 23 %

61. Module mismatch value loss is _____
   a) 3 %
   b) 8 %
   c) 10 %
   d) 12 %

62. The output power, cell efficiency and voltage of a solar cell _____ if its operates at temperature above 25°C
   a) Reduces
   b) Remains Same
   c) Increases
   d) None

63. Manufacture tolerance level of output power is _____
   a) ± 3 %
   b) ± 8 %
   c) ± 9 %
   d) ± 12 %

64. PV module output is reduced by _____
    a) Manufacturing Tolerance   b) Module Mismatch Loss
    c) Option a & b   d) None

65. The actual power output from a PV module is _____ than the maximum power point
    a) Lower   b) Higher
    c) Option a & b   d) None

66. Electricity generation from coal energy in India is _____
    a) 55 %   b) 75 %
    c) 85 %   d) 90 %

67. By adopting solar thermal technology, the solar radiation energy is converted into _____ Energy
    a) Heat   b) Electrical
    c) Mass   d) Chemical

68. In PV modules power rating symbol of "Wp ", subscript "p" refers to _____
    a) Peak   b) Power
    c) Panel   d) None

69. Several PV modules is connected together is known as _____
    a) PV Array   b) PV Node
    c) PV Mesh   d) PV Network

70. Higher value of fill factor is the _____ solar module
    a) Better   b) Worst
    c) Option a & b   d) None

71. A solar cell is a _____ device
    a) Semi - conductor   b) Non conducting
    c) Insulator   d) None

72. A solar is a _____ terminal power generating device
    a) Two
    b) Three
    c) Four
    d) None

73. Many solar cells are connected together to form solar _____
    a) PV Modules
    b) PV Array
    c) Option a & b
    d) None

74. String modules are connected in _____
    a) As per customer requirement
    b) Parallel
    c) Series
    d) Series and Parallel

75. Several solar PV modules are connected together to make _____
    a) PV Array
    b) PV Module
    c) Option a & b
    d) None

76. Voltage of PV Module _____ if temperature rises
    a) Decreases
    b) Increases
    c) Remains Same
    d) None

77. The parameters used in the PV module is _____
    a) Short circuit current (Isc)
    b) Open circuit voltage (Voc)
    c) Current at maximum power point (I m)
    d) Voltage at maximum power point (Vm)
    e) All of these

78. Open circuit voltage of panel is the _____ available from a solar panel
    a) Maximum Voltage
    b) Minimum Voltage
    c) Option a & b
    d) None

79. Short circuit current of panel is the _____ available from a solar panel
    a) Maximum Current
    b) Minimum Current
    c) Option a & b
    d) None

80. _____ is used to generate electricity in PV system
    a) Light
    b) Heat
    c) Temperature
    d) Wind

81. _____ parameter is available in PV module
    a) Fill Factor
    b) Efficiency
    c) Maximum power point
    d) All of these

82. The maximum power will be delivered from the module, when _____
    a) Product of current and voltage at maximum power point
    b) Product of short circuit current and short circuit voltage
    c) Product of open circuit voltage and open circuit current
    d) None

83. "Knee" point of the I-V curve refers to _____
    a) Maximum Power Output
    b) Minimum Power Output
    c) Option a & b
    d) None

84. Short circuit current also known as _____
    a) Maximum Current
    b) Minimum Current
    c) Option a & b
    d) None

85. _____ is known open circuit voltage
    a) Maximum Voltage
    b) Minimum Voltage
    c) Option a & b
    d) None

86. Maximum current (Im) is always _____ than the short circuit current (Isc)
    a) Lower
    b) Higher
    c) Same
    d) None

87. Maximum voltage (Vm) is always _____ than the open circuit voltage (Voc)
    a) Lower
    b) Higher
    c) Same
    d) None

88. _____ Ratio is relevant to solar PV system
    a) Performance
    b) Actual
    c) Design
    d) Project

89. In solar PV system, PR means _____
    a) Performance Ratio
    b) Poor Report
    c) Performance Report
    d) Personal Report

90. Bypass diode has overcome the effect of _____
    a) Short Circuit
    b) Shading
    c) Temperature Rise
    d) Over Voltage

91. PV Module generates electricity in _____
    a) Direct Current
    b) Distribution Current
    c) Alternating Current
    d) None

92. The purpose of blocking diode is to _____
    a) Block the reverse current
    b) Send the excess current to grid
    c) Block the reverse power flow
    d) Protect the equipment against lightning

93. Input power of inverter is indirectly depends on _____
    a) Loss
    b) Output Power
    c) Efficiency
    d) None

94. The battery contains _____ acid
    a) Sulfuric
    b) Hydrochloric
    c) Gastric
    d) None

95. Battery design is depends on _____
    a) Backup Hour
    b) Voltage
    c) Design Load
    d) All of these

96. If battery is 12 V connected, then the PV module voltage level is _____
    a) 9 V
    b) 10 V
    c) 12 V
    d) 15 V

97. _____ is the standard battery voltage level
    a) 10.39 V
    b) 12.0 V
    c) 13.2 V
    d) 14.4 V

98. The purpose of anti-islanding protection for a grid connected inverter is to _____
    a) Avoid reverse power flow during grid failure
    b) Disconnect power from battery
    c) Disconnect from net meter
    d) Disconnect from transformer

99. _____ factor indicates battery discharge level
    a) DOD
    b) Ampere Hour
    c) Capacity
    d) Voltage

100. Cable overheating results in _____
 a) Fire
 b) Temperature Rise
 c) Heat Dissipation
 d) All of these

101. _____ factor is determined the efficient of PV plant
 a) Performance Ratio
 b) Location of Plant
 c) Connected Load
 d) Labour Strength

102. PV module voltage is always _____ than battery voltage
 a) Higher
 b) Lower
 c) Remains same
 d) None

Chapter 6

Pre & Post Commissioning Inspection of the Grid Connected Rooftop SPV Power Plant

1. The things to be taken care before commissioning of plant is _____

   a) No vegetation
   b) Uniform leveling of land
   c) Option a & b
   d) None

2. The possibility of problem occur while doing mechanical and visual inspections for PV array area is _____
   a) Loose connection
   b) Panel scratch
   c) Damaged MC4 connector
   d) All of these

3. The possibility of problem occur while doing mechanical and visual inspections of inverter is _____
   a) Noise level is high
   b) Fan is not working
   c) Inverter support structure is loose
   d) All of these

4. The possibility of problem occur while doing mechanical and visual inspection for distribution board is _____
   a) Number of DC string is less connected than actual design
   b) Tag of DC strings are not mentioned
   c) DC cables are not connected with gland
   d) All of these

5. The equipment to be tested after commissioning is _____
   a) PV Module, Battery
   b) Inverter, DC/AC Box
   c) Transformer
   d) All of these

6. _____ inspection is required for PV plant
   a) Ensure Reliability
   b) Minimize Risks
   c) Gain Certainty
   d) All of these

7. Visual inspection of inverter to check the _____
   a) Cabling
   b) Connecting Box
   c) Mounting Location
   d) All of these

8. In Delhi actual operating temperature of the solar cell is 40 °C. The output voltage (V) of a solar cell at STC is 0.5 V. The output voltage decreases by 3 mV/°C. The new value of output voltage after commissioning is _____

   Formula : Reduced open circuit voltage (Voc) = Output voltage of a solar cell at STC - output voltage decreases per degree (Operating temperature - 25 °C)
   Note: Voc (New Reduced Voltage) = $0.5 - 3 \times 10^{-3} (40 - 25)$
   a) 0.45
   b) 0.53
   c) 0.69
   d) 0.79

9. In Delhi actual operating temperature of the solar cell is 40 °C. The output efficiency (η) of a solar cell at STC is 15 %. The output efficiency decreases by – 0.5 %/°C. The new value of output efficiency η after commissioning is _____
   Formula: Reduced efficiency (η) = Output efficiency of a solar cell at STC – output efficiency decreases per degree (Operating temperature – 25 °C)

Note: η (New Reduced efficiency) = 15 – 0.5 x 10⁻² (40 – 25)
a) 11.45 b) 12.65
c) 13.45 d) 14.92

10. In Chennai actual operating temperature of the solar cell is 40 °C. The output power (P) of a solar cell at STC is 2 W. The output power decreases by 0.5 %/°C. The new value of output power (P) after commissioning is _____
Formula : Reduced Power (P) = Output power of a solar cell at STC – output power decreases per degree (Operating temperature – 25 °C)
Note : P (New Reduced Power) = 2 – 0.5 x 10⁻² (40 – 25)
a) 1.45 b) 1.92
c) 2.00 d) 2.50

11. Solar panel has the value of 20 V, 10 A. Aluminium wire data is referred in below table. The voltage drop before pre commissioning of panel is _____

Note : $V_{drop} = I R$

Diameter, d (mm)	Area, A (mm²)	Resistance of Wire R (Ω)
5	10	0.25

a) 2.5 V b) 5 V
c) 10 V d) 25 V

12. Solar panel has the value of 20 V, 10 A. Aluminum wire data is referred in below table. The percentage voltage drop before pre commissioning of panel is _____
Note : % $V_{drop} = (V_{drop} / V \times 100)$

Diameter, d (mm)	Area, A (mm²)	Resistance of Wire R (Ω)
5	10	0.25

a) 5 b) 10
c) 12.5 d) 25

13. Solar panel has the value of 10 V, 10 A. Aluminum wire data is referred in below table. The voltage drop before pre commissioning of panel is _____

Note: $V_{drop} = IR$

Diameter, d (mm)	Area, A (mm²)	Resistance of Wire R (Ω)
5	10	0.5

a) 2.5 V
b) 5 V
c) 10 V
d) 15 V

14. Solar panel has the value of 10 V, 10 A. Aluminum wire data is referred in below table. The percentage voltage drop before pre commissioning of panel is _____

Note: $\% V_{drop} = (V_{drop} / V \times 100)$

Diameter, d (mm)	Area, A (mm²)	Resistance of Wire R (Ω)
5	10	0.5

a) 2.5
b) 10
c) 15
d) 20

15. The power plant capacity before commissioning of the plant is _____

Module Watts	Number of modules per string	Total number of string
250	20	10

Formula: Capacity per string = Number of modules per string × Module Watts
Plant Capacity = Capacity per string × Total number of string

a) 15 KW
b) 50 KW
c) 75 KW
d) 100 KW

16. The total number of modules before commissioning of the plant is _____

Module Watts	Number of modules per string	Total number of string
250	20	10

    Formula : Total modules = Number of modules per string x Total number of string
    a) 150
    b) 170
    c) 200
    d) 300

17. The capacity per string before commissioning of the plant is _____

Module Watts	Number of modules per string	Total number of string
250	20	10

    Formula : Capacity per string = Number of modules per string x Module Watts
    a) 1 KW
    b) 3 KW
    c) 5 KW
    d) 15 KW

18. The total voltage before commissioning of the plant is _____

Module Voltage V	Module Current Amp	Number of modules per string	Total number of string
30	5	20	10

    Formula : Total Voltage = Number of modules per string x Module Voltage
    a) 600 V
    b) 300 V
    c) 150 V
    d) 1 KV

Solar Assessment

19. The total current before commissioning of the plant is _____

Module Voltage V	Module Current Amp	Number of modules per string	Total number of string
30	5	20	10

Formula : Total Current = Total number of String x Module Current
a) 50 A b) 30 A
c) 15 A d) 13 A

20. The total power before commissioning of the plant is _____

Module Voltage V	Module Current Amp	Number of modules per string	Total number of string
30	5	20	10

Formula : Total Power = Total Voltage x Total Current
a) 3 KW b) 8 KW
c) 13 KW d) 30 KW

21. The energy generation after commissioned the PV plant is _____

Module Voltage (V)	Module Current (Amp)	Number of modules per string	Total number of string	Hours of operation
30	5	20	10	1000

Formula : Total energy generation = Total Power x Numbers of hours
a) 30 MWh b) 30 kWh
c) 80 kWh d) 130 MWh

22. Two strings of 2 kWp systems are installed at 2 BHK house. After commissioning, one of the string showed the current value of zero amps. The problem occurred is _____
a) Open Circuit b) Short Circuit
c) Phase Unbalance d) None

23. Two strings of 2 kWp systems are installed at 2 BHK house. After commissioning, one of the strings showed the voltage value of zero amps. The problem occurred is _____
    a) Short Circuit
    b) Open Circuit
    c) Phase Unbalance
    d) None

24. During inspection of two battery voltages showed the value of 1 V & 3 V. The total battery voltage (V), if connected in series is _____
    a) 3
    b) 4
    c) 5
    d) 6

25. During inspection of three battery voltages showed the value of 1 V, 3 V & 3 V. The total battery voltage (V), if connected in series is _____
    a) 3
    b) 6
    c) 7
    d) 8

26. During inspection of three battery currents showed the value of 1 A, 3 A & 3 A. The total battery current (I), if connected in parallel is _____
    a) 6
    b) 7
    c) 8
    d) 9

27. During inspection of two battery currents showed the value of 1 A & 3 A. The total battery current (I), if connected in parallel is _____
    a) 2
    b) 3
    c) 4
    d) 5

28. Three battery voltages, currents of 1 V/2 A, 3 V/2 A & 3 V/2 A are connected in series. The total battery voltage (V) / Current (I) is _____
    a) 7 V / 2 A
    b) 8 V / 3 A
    c) 7 V / 6 A
    d) 3 V / 4 A

29. Three battery voltages, currents of 3 V / 2 A, 3 V / 2 A & 3 V / 2 A are connected in parallel. The total battery voltage (V) / Current (I) is _____

    a) 3 V / 6 A　　　　　　　　　　b) 9 V / 6 A
    c) 6 V / 6 A　　　　　　　　　　d) 3 V / 6 A

30. Three battery currents of 2 A, 6 A & 2 A are connected in parallel. The total battery Current (I) is _____
    a) 10 A　　　　　　　　　　　　b) 8 A
    c) 4 A　　　　　　　　　　　　 d) 2 A

31. Three battery currents of 6 A, 6 A & 6 A are connected in series. The total battery Current (I) is _____
    a) 2 A　　　　　　　　　　　　 b) 6 A
    c) 12 A　　　　　　　　　　　　d) 18 A

32. Three battery voltages of 8 V, 8 V & 8 V are connected in parallel. The total battery voltage (V) is _____
    a) 4 V　　　　　　　　　　　　 b) 8 V
    c) 18 V　　　　　　　　　　　　d) 24 V

33. Three battery voltages of 8 V, 8 V & 8 V are connected in series. The total battery voltage (V) is _____
    a) 24 V　　　　　　　　　　　　b) 8 V
    c) 4 V　　　　　　　　　　　　 d) 2 V

34. A 5 V, 100 AH sealed lead acid battery is discharging time of 20 hours. The current is _____
    a) 5 A　　　　　　　　　　　　 b) 10 A
    c) 12 A　　　　　　　　　　　　d) 15 A

35. A 12 V, 60 AH sealed lead acid battery is discharging time of 5 hours. The current is _____
    a) 2A　　　　　　　　　　　　　b) 12 A
    c) 15 A　　　　　　　　　　　　d) 22 A

36. A 5 V, 20 Ah lead acid battery. The number of batteries required to get voltage of 20 V is _____
    a) 4
    b) 5
    c) 6
    d) 7

37. A single 6 V, 20 Ah lead acid battery. The number of batteries required to get final voltage of 30 V is _____
    a) 5
    b) 6
    c) 7
    d) 8

38. A single 6 V, 20 Ah lead acid battery. The capacity and number of batteries required to get final voltage of 30 V is _____
    a) 20 Ah, 5 numbers
    b) 80 Ah, 4 numbers
    c) 20 Ah, 3 numbers
    d) 80 Ah, 2 numbers

39. A single 10 V, 200 Ah lead acid battery. The capacity and number of batteries required to get final voltage of 50 V is _____
    a) 200 Ah, 5 numbers
    b) 1000 Ah, 4 numbers
    c) 200 Ah, 3 numbers
    d) 800 Ah, 2 numbers

40. A single 10 V, 200 Ah lead acid battery. The capacity in Ah, energy stored in kWh and number of batteries required to get final voltage of 50 V is _____
    a) 200 Ah, 1 kWh, 5 numbers
    b) 1000 Ah, 1 kWh, 4 numbers
    c) 200 Ah, 2 kWh, 3 numbers
    d) 800 Ah, 1 kWh, 2 numbers

41. A single 10 V, 200 Ah lead acid battery. The energy stored in kWh is _____
    a) 1 kWh
    b) 2 kWh
    c) 3 kWh
    d) 4 kWh

42. A single 100 V, 50 Ah lead acid battery. The energy stored in kWh is
   _____
   a) 1 kWh
   b) 3 kWh
   c) 4 kWh
   d) 5 kWh

43. A single 100 V, 50 Ah lead acid battery is discharging time of 5 hours. The power in kW is _____
   a) 1 kW
   b) 2 kW
   c) 3 kW
   d) 4 kW

44. A single 200 V, 1000 Ah lead acid battery is discharging time of 10 hours. The power in kW is _____
   a) 2 kW
   b) 4 kW
   c) 10 kW
   d) 20 kW

45. Four numbers of 20 Ah battery capacity are connected in parallel. The total capacity is _____
   a) 80 Ah
   b) 60 Ah
   c) 20 Ah
   d) 5 Ah

46. Four numbers of 20 Ah battery capacity are connected in series. The total capacity is _____
   a) 5 Ah
   b) 20 Ah
   c) 60 Ah
   d) 80 Ah

47. A single 5 V, 5 Ah lead acid battery. The type of connection and number of batteries required to get final voltage of 20 V is
   _____
   a) 4 numbers of series connections
   b) 4 numbers of parallel connections
   c) 5 numbers of series – parallel connections
   d) 5 numbers of parallel – series connections

48. A single 5 V, 5 Ah lead acid battery. The type of connection and number of batteries required to get final capacity of 20 Ah is _____

a) 4 Numbers of parallel connections
b) 4 Numbers of series connections
c) 5 Numbers of series – parallel connections
d) 5 Numbers of parallel – series connections

49. A single 5 V, 5 Ah lead acid battery. The type of connection and number of batteries required to get final voltage of 20 V, capacity 20 Ah is _____
a) 4x4 Numbers of series and parallel combination
b) 2x4 Numbers of series and parallel combination
c) 4x3 Numbers of series and parallel combination
d) 4x5 Numbers of series and parallel combination

50. A single 2 V, 8 Ah lead acid battery. The type of connection and number of batteries required to get final voltage of 6 V, capacity 16 Ah is _____
a) 3x2 Numbers of series and parallel combination
b) 2x3 Numbers of series and parallel combination
c) 4x3 Numbers of series and parallel combination
d) 4x5 Numbers of series and parallel combination

51. A single 1 V, 3 Ah lead acid battery. The type of connection and number of batteries required to get final voltage of 5 V, capacity 12 Ah is _____
a) 5x4 Numbers of series and parallel combination
b) 4x5 Numbers of series and parallel combination
c) 4x3 Numbers of series and parallel combination
d) 4x5 Numbers of series and parallel combination

52. A single 2 V, 5 Ah lead acid battery. The type of connection and number of batteries required to get final voltage of 10 V, capacity 15 Ah is _____
   a) 5x3 Numbers of series and parallel combination
   b) 3x4 Numbers of series and parallel combination
   c) 4x3 Numbers of series and parallel combination
   d) 4x5 Numbers of series and parallel combination

53. 100 Ah battery has depth of discharge of 60 %. The present capacity is _____
   a) 40 Ah   b) 50 Ah
   c) 60 Ah   d) 80 Ah

54. 100 Ah battery has depth of discharge of 70 %. The present capacity is _____
   a) 30 Ah   b) 50 Ah
   c) 70 Ah   d) 80 Ah

55. 200 Ah battery has depth of discharge of 70 %. The present capacity is _____
   a) 50 Ah   b) 70 Ah
   c) 80 Ah   d) 140 Ah

56. 100 watts of power is input and the output is 90 watts. The power lost during conversion is _____
   a) 8   b) 9
   c) 10   d) 12

57. 100 watts of power is input and the output is 80 watts. The power lost during conversion is _____
   a) 12   b) 18
   c) 19   d) 20

58. PV module has wattage, voltage of 300 W, 12 V. The current in PV module is _____
   a) 20 A
   b) 25 A
   c) 30 A
   d) 35 A

59. The individual current in Amps is _____

Total Loads		
Equipment's	Voltage in V	Load Watts
Fan	12	60
Light	6	12

   a) 2, 5
   b) 4, 2
   c) 5, 2
   d) 5, 3

60. The individual current in Amps is _____

Total Loads		
Equipment's	Voltage in V	Load Watts
Computer	12	72
Exhaust Fan	12	60

   a) 4, 6
   b) 5, 6
   c) 6, 4
   d) 6, 5

61. The total energy consumption in Wh is _____

Equipment	Load (watts)	Hours
Computer	60	2
Exhaust Fan	40	3

   a) 102
   b) 120
   c) 240
   d) 420

62. The individual Ah capacity is _____

Equipment	Energy (Wh)	Voltage (V)
Computer	120	6
Exhaust Fan	40	10

a) 4, 20　　　　　　　　　　b) 15, 3
c) 20, 4　　　　　　　　　　d) None

63. The actual capacity in Ah is _____

Equipment	Required capacity (Ah)	DoD (%)
Battery	100	50

a) 200　　　　　　　　　　b) 120
c) 80　　　　　　　　　　　d) 70

64. The actual capacity in Ah is _____

Equipment	Required capacity (Ah)	DoD (%)
Battery	150	60

a) 250　　　　　　　　　　b) 120
c) 80　　　　　　　　　　　d) 70

65. A school has required the total battery capacity of 300 Ah. One battery size is 30 Ah. The number of batteries is required is _____
Formula : Total number of batteries = Total Ah capacity required / Ah capacity of one battery
a) 10　　　　　　　　　　b) 20
c) 30　　　　　　　　　　d) 40

66. A college has required the total battery capacity of 500 Ah. One battery size is 50 Ah. The number of batteries is required is _____
Formula : Total number of batteries = Total Ah capacity required / Ah capacity of one battery
a) 10　　　　　　　　　　b) 20
c) 30　　　　　　　　　　d) 40

Chapter 6 | 79

67. A factory has required the total battery capacity of 1000 Ah. One battery size is 50 Ah. The number of batteries is required is _____
Formula : Total number of batteries = Total Ah capacity required / Ah capacity of one battery
a) 20
b) 30
c) 40
d) 50

68. Take one 12 V lead acid battery, _____ batteries required to get the final voltage of 60 V
a) 5
b) 8
c) 10
d) 12

69. A DC array has 100 V / 20 A. The DC power produced from PV array is _____
a) 1 KW
b) 2 KW
c) 3 KW
d) 5 KW

70. A DC array has produced 2000 W. If the Power loss is 10 %, then power loss is _____
a) 50 W
b) 100 W
c) 200 W
d) 300 W

71. A DC array has 100 V / 20 A. Power loss is 10 %. The AC power fed to grid is _____
Formula : DC Power produced from PV array – Power Lost in converting DC to AC power
a) 2000 W
b) 1800 W
c) 1500 W
d) 1200 W

72. A DC array has 100 V / 20 A. Power loss is 10 %. If voltage of 200 V AC, the AC current fed into the grid is _____
Formula : AC current fed into grid = AC power / AC voltage
a) 9 A
b) 10 A
c) 11 A
d) 12 A

Solar Assessment

73. The number of module connected in series is _____

Total string voltage (V)	Module Voltage (V)
144	12

Formula : Number of module = Total String voltage / Module Voltage
a) 10   b) 12
c) 14   d) 16

74. If DC current of series connected solar PV string value is 8A, the DC current of individual solar PV module is _____
a) 8 A   b) 9 A
c) 10 A   d) 12 A

75. If DC current of individual solar PV module value is 8 A, the DC current of series connected solar PV string is _____
a) 5 A   b) 6 A
c) 8 A   d) 13 A

76. The DC Power is _____

DC voltage of series-connected solar PV string (V)	DC current of series-connected solar PV string (Amp)
12	5

a) 70 W   b) 60 W
c) 50 W   d) 30 W

77. A solar PV string is rated for Vm = 100 V and Im = 5 A. Design a series - parallel connected solar PV array to generates 5 kW DC power. The DC output voltage of the array is _____

Formula: DC voltage of series-parallel connected solar PV array = DC voltage of series connected string
a) 5 V
b) 100 V
c) 150 V
d) 200 V

78. A solar PV string is rated for Vm = 100 V and Im = 5 A. Design a series-parallel connected solar PV array to generate 5 kW DC power. The DC output current of the array is _____
Formula: Current = Power / Voltage
a) 5 A
b) 10 A
c) 15 A
d) 20 A

79. If 300 W panel is used, the total number of modules required for 100 KW system is _____
a) 250
b) 300
c) 400
d) 500

80. Assume 5 units generated per KW, _____ units is generated for 20 KW plant
a) 100
b) 150
c) 170
d) 200

81. Before the installation of solar PV module one should check the _____
a) Quality of the all PV Components
b) Site Survey Check List
c) Option a & b
d) None of the above

82. _____ should be asked after commissioning of PV plant
a) Maintenance Manuals
b) Operating Manuals
c) Project Report
d) All of these

Chapter 7

Test and Commission Solar PV System

1. A single row of series connected PV modules is referred as _____
   a) String
   b) Ring
   c) Wing
   d) Node

2. Series connected PV string _____
   a) Increases Voltage
   b) Decreases Voltage
   c) Decreases Current
   d) Increases Current

3. Parallel connected PV string _____
   a) Increases Voltage
   b) Decreases Voltage
   c) Decreases Current
   d) Increases Current

4. If panel is completely shaded, the output is _____
   a) Zero
   b) Maximum
   c) Minimum
   d) None

5. The open circuit voltage (Voc) is measured when _____
   a) Isc is equal to Zero
   b) Isc is equal to Maximum
   c) Isc is equal to Infinity
   d) None

6. The voltage in a short circuit is _____
   a) Zero
   b) Minimum
   c) Maximum
   d) None

7. The current in open circuit is _____
   a) Zero
   b) Minimum
   c) Maximum
   d) None

8. The short circuit current (Isc) is measured when _____
   a) Voc is equal to Zero
   b) Voc is equal to Maximum
   c) Voc is equal to Infinity
   d) None

9. The maximum working voltage indicates is to know _____
   a) The PV system capacity
   b) How much risk is involved
   c) More importance of Human Safety
   d) All of these

10. The important of maximum working voltage is _____
    a) Voltage at which the panel produces maximum output power
    b) Voltage at which the panel produces minimum output power
    c) Option a & b
    d) None

11. If two 0.5 V solar cells are connected in series then total voltage value is _____
    a) 1
    b) 2
    c) 3
    d) 4

12. If two 36 V solar cells are connected in series then total voltage value is _____
    a) 12
    b) 27
    c) 72
    d) 82

13. One solar cell value of 0.5 V, _____ solar cells to be connected in series to get a final voltage of 30 V
    a) 5
    b) 6
    c) 50
    d) 60

14. One PV module has the voltage of 20 V, _____ solar cells to be connected in series if cell voltage is 0.25 V
    a) 8
    b) 18
    c) 80
    d) 90

15. If 0.5 V of 10 number of solar cell has connected in series, then the total voltage value is _____
    a) 0.5
    b) 2.5
    c) 5
    d) 5.2

16. If 250 watts solar module is installed, _____ unit produces per day
    a) 1
    b) 2
    c) 3
    d) 4

17. The value of maximum voltage (Vm) is 3 V and short circuit current (Isc) is 2 A, then the maximum power (Pm) value is _____
    Note : Im = 0.90 Isc, Pm = Vm x Im
    a) 5.4
    b) 6
    c) 8
    d) 9

18. The value of maximum voltage (Vm) is 5 V and short circuit current (Isc) is 2 A. The maximum power (Pm) value is _____
    Note : Im = 0.90 Isc, Pm = Vm x Im
    a) 9 W
    b) 6 W
    c) 8 W
    d) 4 W

19. The value of maximum current (Im) is 3 A and open circuit voltage (Voc) is 5 V. The maximum power (Pm) value is _____
    Note : Vm = 0.80 Voc, Pm = Vm x Im
    a) 12 W
    b) 16 W
    c) 18 W
    d) 19 W

20. If PV module generates output of 5 V, 1 A, then type of connection to get an output of 1 Amps and 20 Volts is _____
    a) Series
    b) Parallel
    c) Parallel - Series
    d) Series - Parallel

21. If PV module generates output of 5 V, 1 A, then number of modules required to get an output of 1 Amps and 20 Volts is _____
    a) 1
    b) 2
    c) 4
    d) 5

22. If PV module generates output of 5 V, 1 A, then the type of connection to get an output of 5 Amps and 5 Volts is _____
    a) Series
    b) Parallel
    c) Parallel - Series
    d) Series - Parallel

23. If PV module generates output of 5 V, 1 A, then the number of modules required to get an output of 5 Amps and 5 Volts is _____
    a) 1
    b) 2
    c) 5
    d) 10

24. The value of short circuit current (Isc) = 4 A, maximum current (Im) = 3.5 A, open circuit voltage (Voc) = 1 V, Fill Factor (FF) = 70 %, then the maximum voltage (Vm) is _____
    a) 0.8 V
    b) 1.2 V
    c) 1.5 V
    d) 1.8 V

25. The value of short circuit current (Isc) = 4 A, maximum current (Im) = 3.1 A, open circuit voltage (Voc) = 1 V, Fill Factor (FF) = 72 %, then the maximum voltage (Vm) is _____
    a) 0.92 V
    b) 1 V
    c) 1.2 V
    d) 1.5 V

26. The value of short circuit current (Isc) = 3 A, maximum voltage (Vm) = 3.1 A, open circuit voltage (Voc) = 1 V, Fill Factor (FF) = 82 %, then the maximum current (Im) _____
    a) 0.82 V
    b) 1.2 V
    c) 1.5 V
    d) 1.8 V

27. The value of short circuit current (Isc) = 4 A, maximum current (Im) = 3.5 A V, open circuit voltage (Voc) = 1 V, maximum voltage (V m) = 0.8 V, then the Fill Factor (FF) is _____
    a) 70 %
    b) 80 %
    c) 90 %
    d) 95 %

28. A bypass diode is connected in _____ with solar cell
    a) Series
    b) Parallel
    c) Series - Parallel
    d) None

29. During normal condition (no shadow) the bypass diode is operated in _____ bias
    a) Forward
    b) Reverse
    c) Forward and Reverse
    d) None

30. A bypass diode connected in parallel but with _____ polarity to solar cells
    a) Forward
    b) Reverse
    c) Forward and Reverse
    d) None

31. The purpose of bypass diode is _____
    a) Avoid hot spots
    b) Formation of hot spots
    c) Option a & b
    d) None

32. The purpose of blocking diode is _____
    a) Prevent Reverse Current Flow
    b) Prevent Leakage Current
    c) Option a & b
    d) None

33. Blocking diode prevents the _____ of battery into the SPV module
    a) Discharging
    b) Overcharging
    c) Option a & b
    d) None

34. Blocking diodes are added in PV system to avoid _____ into the PV modules
    a) Reverse Flow of Current
    b) Reverse Voltage
    c) Reverse Power
    d) None

35. Blocking diodes are connected in _____ of PV system
    a) Series
    b) Parallel
    c) Series – Parallel
    d) None of these

36. The process of electrical energy is consumed from battery is called _____ process
    a) Charge
    b) Discharge
    c) Option a & b
    d) None

37. The process of electrical energy is stored in battery is called _____ process
    a) Charge
    b) Discharge
    c) Option a & b
    d) None

38. One time battery is known as _____
    a) Non - Rechargeable Battery
    b) Rechargeable Battery
    c) Option a & b
    d) None

39. The parameter available in battery system is _____
    a) Voltage
    b) Ampere Hour
    c) DOD
    d) All of these

40. Battery parameter is _____
    a) Battery terminal Voltage
    b) Charge storage Capacity
    c) Depth of Discharge
    d) All of these

41. Battery parameters is _____
    a) Charging - Discharging cycle
    b) Life cycle
    c) Self - Discharge
    d) All of these

42. The unit of charge storage capacity in battery is _____
    a) Ampere Hour   b) Voltage
    c) Ampere   d) None

43. Battery is disconnected from the load when it reaches below is _____
    a) Cut Off Voltage   b) Cut On voltage
    c) Cut In - Out Voltage   d) None

44. Capacity of battery depends on _____
    a) Temperature   b) Volume
    c) Pressure   d) None

45. DOD means _____
    a) Depth of Discharge   b) Drain of Discharge
    c) Discharge of Depth   d) Design of Discharge

46. SOC means _____
    a) State of Charge   b) Stand of Charge
    c) Step of Charge   d) Sleep of Charge

47. SoC has decreased while battery voltage _____
    a) Decreases
    b) Increases
    c) Remains Same
    d) None

48. SoC is directly proportional to the _____
    a) Battery Open Circuit Voltage
    b) Inverter open circuit voltage
    c) Rectifier Open Circuit Voltage
    d) None

49. Over discharge of battery has affected the _____
    a) Life of battery
    b) Capacity
    c) Mechanical Damage
    d) All of these

50. C Rating is related to _____
    a) Battery
    b) Rectifier
    c) Inverter
    d) None

51. _____ unit is used for C Rating
    a) Ampere
    b) Voltage
    c) Time
    d) None

52. C / 1 or 1 C of battery refers _____
    a) C - Capacity, 1 hour for full charge or discharge
    b) C - Current, 1 hour for full charge
    c) C - Continuous, 1 hour for full discharge
    d) C - Charge, 1 hour for full charge or discharge

53. Battery terminal voltage reduces when the temperature _____
    a) Increases
    b) Decreases
    c) Remains Same
    d) None

54. One Charing and One discharging is called as _____ cycle of the battery
    a) One
    b) Two
    c) Three
    d) Four

55. The two main parameters of battery are _____
    a) Voltage
    b) Ampere Hour
    c) Option a & b
    d) None

56. Voltage _____ when batteries are connected in series
    a) Increases
    b) Decrease
    c) Option a & b
    d) None

57. Current _____ when batteries are connected in series
    a) Increases
    b) Decreases
    c) Remains Same
    d) None

58. Voltage _____ when batteries are connected in parallel
    a) Increases
    b) Decreases
    c) Remains Same
    d) None

59. Current _____ when batteries are connected in parallel
    a) Increases
    b) Decreases
    c) Remains Same
    d) None

60. Capacity is _____ when batteries connected in parallel
    a) Added
    b) Subtracted
    c) Divided
    d) Remains Same

61. Capacity is _____ when batteries connected in series
    a) Added
    b) Subtracted
    c) Divided
    d) Remains Same

62. Inverter output power has increased while efficiency is _____
    a) Increases
    b) Decreases
    c) Remains Same
    d) None

63. Output power of inverter is directly depends on the _____
    a) Efficiency
    b) Input Power
    c) Option a & b
    d) None

64. Central inverter is used for _____
    a) 250 KW - 1 MW
    b) 10 KW - 50 KW
    c) Option a & b
    d) None

65. Voltage range of central inverter is _____
    a) 400 - 1 KV
    b) 100 - 200 V
    c) Option a & b
    d) None

66. Central inverter is suitable for identical _____
    a) Type
    b) Make
    c) Power Rating
    d) All of these

67. String inverter is suitable for non - identical _____
    a) Type
    b) Make
    c) Power Rating
    d) All of these

68. Voltage range of string inverter is _____
    a) 200 - 600 V
    b) 1 KV - 2 KV
    c) Option a & b
    d) None

69. _____ is used in each module
    a) Module / Micro Inverter
    b) Central Inverter
    c) String Inverter
    d) None

70. 100 watts of power is input and the output is 92 watts, then the inverter efficiency is _____
   a) 80 %
   b) 92 %
   c) 95 %
   d) 98 %

71. 100 watts of power is input and the output is 95 watts, then the inverter efficiency is _____
   a) 89 %
   b) 90 %
   c) 95 %
   d) 98 %

72. 1000 watts of power is input and the output is 950 watts, then the inverter efficiency is _____
   a) 90 %
   b) 95 %
   c) 98 %
   d) 99 %

73. 100 watts of power is input and the output is 96 watts, then the inverter efficiency and percentage of power lost during conversion is _____
   a) 92 %, 8 %
   b) 95 %, 5 %
   c) 96 %, 4 %
   d) 98 %, 2 %

74. 100 watts of power is input and the output is 98 watts, then the percentage of power lost during conversion is _____
   a) 2 %
   b) 3 %
   c) 5 %
   d) 8 %

75. An inverter input power of 300 watts and efficiency is 92 %, then the output power is _____
   a) 236
   b) 267
   c) 276
   d) 326

76. An inverter input power of 200 watts and efficiency is 90 %, then the output power is _____
   a) 108
   b) 180
   c) 190
   d) 200

77. An inverter output power of 150 watts and efficiency is 50 %, then the input power is _____
   a) 150
   b) 200
   c) 300
   d) 350

78. An inverter output power of 100 watts and efficiency is 50 %, then the input power is _____
   a) 150
   b) 200
   c) 300
   d) 350

79. If voltage is 20 (Vrms) and resistance R is 5 kΩ, the RMS AC current (Irms) is _____
   a) 2A
   b) 4A
   c) 8A
   d) 12A

80. The unit of resistivity is _____
   a) Ohm Meter
   b) Ohm
   c) Meter
   d) None

81. The unit of resistance is _____
   a) Ohm
   b) Farad
   c) Henry
   d) None

82. If we increase the cross section of the area, it will lower the _____ value
   a) Resistance
   b) Resistivity
   c) Option a & b
   d) None

83. If we decrease the cross section of the area, it will higher the _____ value
    a) Resistance
    b) Resistivity
    c) Option a & b
    d) None

84. Low resistance of the wire gives the _____ voltage drop
    a) Lower
    b) Higher
    c) Option a & b
    d) None

85. Which one of the following is correct ?
    a) Temperature ∝ Resistance
    b) Temperature ∝ 1 / Resistance
    c) Temperature ∝ Durability
    d) None

86. Which one of the following is correct for selecting wire ?
    a) Temperature ∝ Voltage Drop
    b) Temperature ∝ 1 / Voltage drop
    c) Temperature ∝ Durability
    d) None

87. In AC circuits only one parameter is different than DC circuit _____
    a) Power Factor
    b) Voltage
    c) Current
    d) None

88. If voltage is 20 Vrms and RMS AC current (Irms) is 5A, then the apparent power (Papp) is _____
    a) 4 W
    b) 15 W
    c) 25 W
    d) 100 W

89. If apparent power is 100 W and Power Factor is 0.8, then the Real Power (Preal) is _____
    a) 8 W
    b) 80 W
    c) 108 W
    d) 800 W

90. A tube light is consuming 1000 units and used for 10 hours, then the connected power is _____
    a) 1 W
    b) 10 W
    c) 100 W
    d) 1000 W

91. If Energy of 200 kWh is utilized for 5 hours, then the power consumed is _____
    a) 15 KW
    b) 25 KW
    c) 30 KW
    d) 40 KW

92. A motor is consuming 15 units per day, then the monthly (30 days) power consumption _____ units
    a) 405
    b) 450
    c) 504
    d) 540

93. An electrical appliance is connected to 230 V which results in 5A current through the load. The power consumed by the load is _____
    a) 1.150 W
    b) 11.05 KW
    c) 11.50 W
    d) 1150 W

94. If two bulbs A and B with same power of 70 W from same manufacturer. Bulb A is used for 2 hours while bulb B is used for 4 hours, then _____ will glow more brightly
    a) Bulb A & B
    b) Bulb A
    c) Bulb B
    d) None

95. If two bulbs A and B with same power of 70 W from same manufacturer. Bulb A is used for 2 hours while bulb B is used for 4 hours, then _____ will consume more energy
    a) Bulb A & B
    b) Bulb A
    c) Bulb B
    d) None

96. A 100 W tube light is running for 8 hrs per day, then the energy consumption per day in kWh is _____
    a) 0.08
    b) 0.8
    c) 8
    d) None

97. If three number of 10 W CFL is running at 8 hrs per day, then the daily energy requirement in Wh is _____
    a) 240
    b) 80
    c) 30
    d) 24

98. If 100 W bulb is used for 2 hours, then the energy consumed is _____
    a) 2 Wh
    b) 50 Wh
    c) 100 Wh
    d) 200 Wh

99. One day consumption of 36 kWh is taking for 6 hrs, then the actual load requirement is _____
    a) 6 KW
    b) 7 KW
    c) 8 KW
    d) 10 KW

100. If Bulb A - 100 W is used for 5 hours and Bulb B - 50 W is used for 10 hours, then the energy consumed is _____
    a) 500 Wh in Bulb A, 500 Wh in Bulb B
    b) 50 Wh in Bulb A, 50 Wh in Bulb B
    b) 50 Wh in Bulb A, 500 Wh in Bulb B
    d) 500 Wh in Bulb A, 50 Wh in Bulb B

101. If present load of 10 KW and 20 % to be added in future load, then the total load requirement is _____
    a) 8 KW
    b) 10 KW
    c) 12 KW
    d) 14 KW

102. A 3 Phase distribution transformer capacity is 100 KVA. As per net metering policy, Distribution Transformer (DT) is only allowed 50 % of the solar power, then the allowable feed in capacity to the transformer is _____
    a) 50 kVA
    b) 55 KVA
    c) 60 KVA
    d) 100 kVA

103. A 3 Phase distribution transformer capacity is 100 KVA. As per net metering policy, Distribution Transformer (DT) is only allowed 90 % of the solar power. Find the allowable feed in capacity to the transformer is _____
    a) 55 KVA
    b) 70 KVA
    c) 90 kVA
    d) 100 kVA

104. Voltage is measured in _____
    a) Voltmeter
    b) Ammeter
    c) Wattmeter
    d) None

105. _____ value of current is showed in digital ammeter
    a) RMS
    b) Peak
    c) Both a & b
    d) None

106. Multimeter is used to measure the _____
    a) Voltage
    b) Current
    c) Resistance
    d) All of these

107. Multimeter can be used to measure the _____
    a) DC
    b) AC
    c) Option a & b
    d) None

108. In electrical circuit, Ammeter is connected in _____
    a) Series
    b) Parallel
    c) Series – Parallel
    d) None

109. In electrical circuit, Voltmeter is connected in _____
    a) Series
    b) Parallel
    c) Series - Parallel
    d) None

110. The negative value of measured voltage in the display of multimeter due to _____ position of the probes
    a) Interchanging
    b) Left
    c) Centre
    d) Right

111. _____ meter is available to measure the power
    a) Wattmeter
    b) Ammeter
    c) Voltmeter
    d) None

112. Wattmeter coil in series measures the _____
    a) Current
    b) Voltage
    c) Power
    d) Energy

113. Wattmeter coil in parallel measures the _____
    a) Current
    b) Voltage
    c) Power
    d) Energy

114. Energy Meter is also known as _____
    a) Watt - Hour Meter
    b) Current Meter
    c) Bill Meter
    d) Voltage Meter

115. Maximum power rating that a PV module can provide under best condition is _____
    a) Standard Test Condition (STC)
    b) Standard Temperature Condition (STC)
    c) Standard Best Condition (SBC)
    d) Standard Terminal Condition (STC)

116. _____ is correct for PV module testing parameters under Standard Test Condition (STC)
   a) 1000 W/m², 1 m/s and 250 °C
   b) 800 W/m², 2 m/s and 150 °C
   c) 1000 W/m² and 250 °C
   d) 900 W/m² and 350 °C

117. Multimeter is used to measure _____ & _____
   a) Voltage, Current
   b) Power, Current
   c) Power, Voltage
   d) Resistance, Power

118. If 30 numbers of 0.5 V of solar cell is connected in _____, then it gives 15 V output
   a) Series
   b) Parallel
   c) Series - Parallel
   d) None of these

119. If 30 numbers of 0.5 V of solar cell is connected in _____, then it gives 0.5 V output
   a) Series
   b) Parallel
   c) Series - Parallel
   d) None of these

120. If 30 numbers of 0.5 A of solar cell is connected in _____, then it gives 15 A output
   a) Series
   b) Parallel
   c) Series - Parallel
   d) None of these

121. If 30 numbers of 0.5 A of solar cell is connected in _____, then it gives the 0.5 A output
   a) Series
   b) Parallel
   c) Series - Parallel
   d) None of these

122. Voltage is increased while connected in series, then the power also _____
   a) Increases
   b) Decreases
   c) Option a & b
   d) None

123. Opposite polarity terminals of PV modules connected together is named as _____
   a) Series
   b) Parallel
   c) Series - Parallel
   d) None of these

124. Same polarity terminals of PV modules connected together is named as _____
   a) Series
   b) Parallel
   c) Series - Parallel
   d) None of these

125. Positive terminal of one module to negative terminal of the next module is named as _____
   a) Series
   b) Parallel
   c) Series - Parallel
   d) None of these

126. If a DC motor works on 12 V and while running it takes 2 A current, then the DC power consumed by the motor is _____
   a) 12 W
   b) 24 W
   c) 32 W
   d) 48 W

127. If a DC motor operates at 220 V and consumes 2200 watt power, then the DC current required to run the motor is _____
   a) 5 Amp
   b) 10 Amp
   c) 15 Amp
   d) 20 Amp

128. In series connection, if Module 1 Voc 1 is 12 V, Module 2 Voc 2 is 12 V and Module 3 Voc 3 is 12 V, then the total open circuit voltage (Voc) is _____
   Formula : Total Voc = Voc 1 + Voc 2 + Voc 3
   a) 12 V
   b) 24 V
   c) 36 V
   d) 48 V

129. In parallel connection, if Module 1 Isc 1 is 2 A, Module 2 Isc 2 is 3 A and Module 3 Isc 3 = 5 A, then the total short circuit current (Isc), _____

Formula : Total Isc = Isc 1 + Isc 2 + Isc 3
a) 5 A
b) 7 A
c) 8 A
d) 10 A

130. In series connection, if Module 1 Voc 1 is 12 V, Isc 1 is 2 A, Module 2 Voc 2 is 12 V, Isc 2 is 2 A and Module 3 Voc 3 is 12 V, Isc 3 is 2 A, then the total open circuit voltage (Voc) and short circuit current (Isc) is _____
a) 12 V, 2 A
b) 24 V, 6 A
c) 36 V, 2 A
d) 36 V, 4 A

131. If Module 1 P 1 is 2 W, Module 2 P 2 is 3 W and Module 3 P 3 is 5 W, then the total Power (P) is _____
a) 2 W
b) 5 W
c) 7 W
d) 10 W

132. In parallel connection, if Module 1 Voc 1 is 12 V, Isc 1 is 2 A, Module 2 Voc 2 is 12 V, Isc 2 is 2 A, and Module 3 Voc 3 is 12 V, Isc 3 is 2 A, then the total power circuit voltage is _____
a) 72 W
b) 36 W
c) 24 W
d) 12 W

133. If Module 1 Voc 1 is 12 V, Isc 1 is 3 A, Module 2 Voc 2 is 12 V, Isc 2 is 2 A, and Module 3 Voc 3 is 12 V, Isc 3 is 5 A, then the individual power of P 1, P 2, P 3 is _____
a) 36 W, 24 W, 60 W
b) 36 W, 60 W, 24 W
c) 24 W, 60 W, 36 W
d) 12 W, 24 W, 36 W

134. If Module 1 Voc 1 is 10 V, Isc 1 is 3 A, Module 2 Voc 2 is 20 V, Isc 2 is 5 A, and Module 3 Voc 3 is 30V Isc 3 is 5 A, then the individual power of P 1, P 2, P 3 is _____
a) 30 W, 100 W, 150 W
b) 100 W, 150 W, 24 W
c) 150 W, 60 W, 30 W
d) 30 W, 24 W, 36 W

135. If Module 1 Voc 1 is 10 V, Isc 1 is 3 A, Module 2 Voc 2 is 20, V Isc 2 is 5 A and Module 3 Voc3 is 30 V, Isc 3 is 5 A, then the total power of P if the three modules are connected independently is
_____
Formula : Total power is equal to sum of individual module power
a) 28 W
b) 36 W
c) 280 W
d) 820 W

136. If Module 1 Voc 1 is 10 V, Isc 1 is 5 A, Module 2 Voc 2 is 20 V, Isc 2 is 5 A and Module 3 Voc3 is 30 V, Isc3 is 3 A, then the total power of P if the three modules are connected independently is _____
Formula : Total power is equal to sum of individual module power
a) 203 W
b) 204 W
c) 208 W
d) 240 W

137. The total power of P is _____, if the three modules are connected in series. Consider "mismatch voltage" of each module Voc 1 is 10 V, Isc 1 is 5 A Module 2 Voc 2 is 3 V, Isc 2 is 5 A, Module 3 Voc 3 is 5 V, Isc 3 is 5 A.
Formula : Total power is equal to sum of individual module power
a) 45 W
b) 54 W
c) 90 W
d) 95 W

138. The total power of P is _____, if the three modules are connected in series. Consider "mismatch current" of each module no.1 Voc 1 is 10 V, Isc 1 is 3 A, Module 2 Voc 2 is 5 V, Isc 2 is 5 A, Module 3 Voc 3 is 5 V, Isc 3 is 6 A.
Formula : Total voltage = Sum of individual module voltage
Total current = Take the lowest current
Total power = Total Voltage x Lowest Current
a) 40 W
b) 54 W
c) 60 W
d) 90 W

139. The total short circuit current Isc is _____, if the two modules are connected in parallel. Consider each module Isc 1 is 3 A, Module Isc 2 is 5 A
   a) 2
   b) 3
   c) 5
   d) 8

140. The total short circuit current Isc is _____, if the three modules are connected in parallel. Consider each Module Isc 1 is 3 A, Module Isc 2 is 5 A, Module Isc 3 is 6 A
   a) 5
   b) 6
   c) 9
   d) 14

141. The total power of P is _____, if the three modules are connected in parallel. Consider "Same voltage" of each Module Voc 1 is 5 V, Isc 1 is 3 A, Module 2 Voc 2 is 5 V, Isc 2 is 5 A, Module 3 Voc 3 is 5 V, Isc 3 is 6 A.
   Formula : Total power is equal to sum of individual module power
   a) 45 W
   b) 55 W
   c) 70 W
   d) 80 W

142. The total short circuit current Isc is _____, if the three modules are connected in parallel. Consider "Same voltage" of each module Voc 1 is 5 V, Isc 1 is 3 A, Module 2 Voc 2 is 5 V, Isc 2 is 5 A, Module 3 Voc 3 is 5 V Isc 3 is 6 A
   Formula : Total short circuit current is equal to sum of individual current
   a) 5
   b) 6
   c) 8
   d) 14

143. The total open circuit voltage Voc is _____, if the three modules are connected in parallel. Consider "Same voltage" of each Module Voc 1 is 5 V, Isc 1 is 3 A, Module 2 Voc 2 is 5 V, Isc 2 is 5 A, Module 3 Voc 3 is 5 V, Isc 3 is 6 A

Formula : Total open circuit voltage is equal to individual module voltage
a) 3 b) 5
c) 6 d) 8

144. The total open circuit voltage Voc is _____, if the three modules are connected in series. Consider "Same voltage" of each Module Voc 1 is 5 V, Isc 1 is 3 A, Module 2 Voc 2 is 5 V, Isc 2 is 5 A, Module 3 Voc 3 is 5 V, Isc 3 is 6 A
Formula : Total open circuit voltage is equal to sum of individual module voltage
a) 5 b) 10
c) 15 d) 20

145. The total short circuit current Isc is _____, if the three modules are connected in series. Consider identical current of each Module Voc 1 is 5 V, Isc 1 is 3 A, Module 2 Voc 2 is 5.1 V, Isc 2 is 3 A, Module 3 Voc 3 is 5.5 V, Isc 3 is 3 A
Formula : Total short circuit current is equal to individual module current
a) 3 b) 10
c) 15 d) 20

146. Battery is manufactured at which standard temperature condition is _____
a) 25 °C b) 35 °C
c) 45 °C d) 55 °C

147. Battery terminal voltage has decreased while DOD of battery is _____
a) Increased b) Decreased
c) Remains Same d) None

148. SOC indicates _____
    a) Charge level of battery
    b) Discharge level of battery
    c) Charge – Discharge level of battery
    d) None

149. A battery capacity of 10 Ah is discharged in 2 hours. The output current is _____
    a) 2
    b) 4
    c) 5
    d) 10

150. A battery capacity of 20 Ah is discharged in 10 hours. The output current is _____
    a) 2
    b) 5
    c) 10
    d) 20

151. A battery capacity of 10 Ah is discharged in 2 hours. The output current is _____
    a) 2
    b) 4
    c) 5
    d) 10

152. 10 V battery is having capacity 120 Ah. The power of the battery is _____, if discharge duration is 10 hours.
    Formula : Power = Voltage x Current, Current = Capacity / discharge duration
    a) 120 W
    b) 110 W
    c) 12 W
    d) 10 W

153. 10 V battery is having capacity 120 Ah. The energy stored (wh) in the battery is _____. if discharge duration is 10 hours.
    Formula : Power = Voltage x Current , Current = Capacity / discharge duration
    Energy (wh) = Voltage x Capacity
    a) 1000
    b) 1010
    c) 1020
    d) 1200

154. 15 V battery is having capacity 150 Ah. The energy stored (kWh) in the battery is _____. if discharge duration is 10 hours.
Formula : Power = Voltage x Current, Current = Capacity / discharge duration
Energy (kwh) = Voltage x Capacity
   a) 2.25   b) 1.25
   c) 1.02   d) 1.01

155. The SoC value is _____, if DoD is 60 %
   a) 40 %   b) 50 %
   c) 60 %   d) 70 %

156. The DoD value is _____, if SoC of 70 %
   a) 30 %   b) 40 %
   c) 50 %   d) 70 %

157. The SoC value is _____, if DoD of 80 %
   a) 20 %   b) 30 %
   c) 40 %   d) 70 %

158. The DoD value is _____, if SoC of 50 %
   a) 40 %   b) 50 %
   c) 60 %   d) 70 %

159. The power of P is _____, if Module of Voc is 12 V, Isc is 3 A
   a) 36 W   b) 24 W
   c) 24 W   d) 12 W

160. The power of P is _____, if Module of Voc is 10 V, Isc is 5 A
   a) 10 W   b) 12 W
   c) 24 W   d) 50 W

161. Repeated charging and discharging of battery is known as _____
    a) Rechargeable Battery
    b) Non-Rechargeable Battery
    c) Option a & b
    d) None

162. Open circuit voltage is measured when circuit is _____
    a) Open
    b) Short
    c) Open - Short
    d) None

163. A 10 Ah battery is discharged at 25 °C temperature, it gives _____
    a) Maximum capacity
    b) Minimum Capacity
    c) Option a & b
    d) None

164. A 10 Ah battery is discharged at 20 °C temperature than STC of 25 °C, it gives _____
    a) Less Capacity
    b) More Capacity
    c) Option a & b
    d) None

165. When the temperature decreases, the battery capacity also _____
    a) Decreases
    b) Increases
    c) Remains same
    d) None

166. At 38 °C the battery shows _____ charge than 20 °C
    a) Higher
    b) Lower
    c) Remains same
    d) None

167. At 20 °C the battery shows _____ charge than 38 °C
    a) Lower
    b) Higher
    c) Remains same
    d) None

168. _____ is deposited on battery contacts
   a) Copper Sulphate
   b) Sodium Sulphate
   c) Sulphur Oxide
   d) Sulphur Phosphate

169. Specific gravity is measured on _____
   a) Battery
   b) Inverter
   c) Rectifier
   d) UPS

170. Specific gravity is measured by _____
   a) Hygrometer
   b) Voltmeter
   c) Ammeter
   d) Multimeter

171. Clamp meter is non - contact _____
   a) Multimeter
   b) Hygrometer
   c) Option a & b
   d) None

172. Clamp Meter is used to measure _____
   a) Current
   b) Voltage
   c) Power
   d) None

173. The device which runs on AC load is _____
   a) Fan
   b) Fridge
   c) Washing Machine
   d) All of these

174. The device which runs AC load is _____
   a) TV
   b) Grinder
   c) Pump
   d) All of these

175. A multimeter is measured by the source and end voltage is 12 V and 11.3 V. The voltage drop is _____
   a) 0.7 V
   b) 0.5 V
   c) 0.8 V
   d) 0.1 V

176. DC system has a voltage, current of 10 V, 10 A, and 5 % voltage drop. The resistance value is _____
Formula : V_{drop} = I R
a) 0.001
b) 0.002
c) 0.005
d) 0.006

177. DC system has a voltage, current of 15 V, 10 A and 5 % voltage drop. The resistance value is _____
Formula : V_{drop} = I R
a) 0.003
b) 0.006
c) 0.008
d) 0.009

178. PV module has a voltage and power of 10 V, 50 W. The current rating is _____
a) 2.5 A
b) 3 A
c) 4 A
d) 5 A

179. The energy requirement in Wh from solar module is _____

Total Energy (kWh)	Battery Efficiency (%)
800	80

a) 1000
b) 1500
c) 2000
d) 2500

180. The energy consumption in Wh is _____

Equipment's	Watts	Number of Equipment's	Hours
Fan	60	2	2
Light	40	2	2

a) 200
b) 300
c) 400
d) 500

181. The energy consumption in Wh is _____

Equipment's	Watts	Number of Equipment's	Hours
Fan	20	2	2
Light	20	2	2

a) 90
b) 100
c) 160
d) 180

182. The total power in KW is _____

Equipment's	Watts
Fan	60
Light	60

a) 90
b) 120
c) 180
d) 170

183. Find the inverter capacity _____

Formula : Inverter Capacity = Load / Inverter Efficiency

Total Load watts	Inverter Efficiency
100	0.8

a) 80 VA
b) 90 VA
c) 100 VA
d) 125 VA

184. The actual battery capacity Ah is _____

Formula : Battery capacity Ah = Energy Units / Battery Voltage

Actual Battery Capacity Ah = Battery capacity Ah / DoD

Energy units in Wh	Battery voltage V	DOD %
300	15	0.8

a) 18
b) 25
c) 20
d) 15

185. The power from PV module is _____

Total Power KW	Battery Efficiency %
300	0.8

a) 375
c) 425
b) 400
d) 500

186. Take one 12 V lead acid battery. _____ connection required to get the final voltage of 60 V
   a) Series
   b) Parallel
   c) Series - Parallel
   d) None

187. Take one 12 Ah lead acid battery. _____ batteries required to get the final voltage of 60 Ah
   a) 5
   b) 8
   c) 10
   d) 12

188. Take one 12 Ah lead acid battery. _____ connection required to get the final voltage of 60 Ah
   a) Series
   b) Parallel
   c) Series - Parallel
   d) None

189. The number of module connected in parallel is _____
   Formula : Number of module = Total String current / Module current

Total string current in Amp	Module current in Amp
150	5

a) 15
c) 30
b) 20
d) 45

190. DC voltage of parallel connected solar PV string value is 8 V. The DC voltage of individual solar PV module is _____
    a) 8 V
    b) 9 V
    c) 10 V
    d) 12 V

191. DC voltage of individual solar PV module value is 10 V. The DC voltage of parallel connected solar PV string is _____
    a) 8 V
    b) 10 V
    c) 13 V
    d) 15 V

192. The DC Power is _____

DC voltage of parallel connected solar PV string in V	DC current of parallel connected solar PV string in Amp
20	8

    a) 130 W
    b) 150 W
    c) 160 W
    d) 170 W

193. The PV energy generated is less than energy consumed by the load, the net energy meter runs in _____
    a) Forward Direction
    b) Reverse Direction
    c) Option a & b
    d) None

194. The PV energy generated is less than energy consumed by the load, the net energy meter shows in _____ reading
    a) Positive
    b) Negative
    c) Option a & b
    d) None

195. The PV energy generated is more than energy consumed by the load, the net energy meter runs in _____
    a) Forward Direction
    b) Reverse Direction
    c) Option a & b
    d) None

196. The PV energy generated is more than energy consumed by the load, the net energy meter shows in _____ reading
    a) Positive
    b) Negative
    c) Option a & b
    d) None

197. Net meter reading is _____ when energy is drawn from grid
    a) Positive
    b) Negative
    c) Option a & b
    d) None

198. Net meter reading is _____ when energy is fed to grid
    a) Positive
    b) Negative
    c) Option a & b
    d) None

199. The PV and load energy meters are _____ and registers the flow of energy in _____ direction
    a) Uni - Directional, Forward
    b) Bi - Directional, Forward
    c) Bi - Directional, Reverse
    d) Uni - Directional, Reverse

200. The net energy meter is _____ and registers the flow of energy in both _____ directions
    a) Bi - Directional, Forward & Reverse
    b) Uni - Directional, Forward & Reverse
    c) Option a & b
    d) None

201. The formula used for PV generated energy is equal to _____
    a) Energy consumed by load + Energy fed to grid
    b) Energy consumed by load - Energy fed to grid
    c) Option a & b
    d) None

202. If PV meter reads 10 kWh and Load meter reads 8 kWh, _____ be the Net energy meter reading
Formula: Energy fed to grid = PV generated energy – Energy consumed by load
a) – 2 kWh
b) + 2 kWh
c) 5 kWh
d) 6 kWh

203. If net energy meter reads + 4 kWh and load meter reads 10 kWh, The PV generated energy is _____
Formula :
PV generated energy = Energy consumed by load – Energy fed to grid
Net meter reading is positive when energy is drawn from grid
a) 5 kWh
b) 6 kWh
c) 8 kWh
d) 12 kWh

204. Performance ratio of PV power plant varies between _____
a) 70 % to 80 %
b) 50 % to 60 %
c) 30 % to 40 %
d) 10 % to 20 %

205. _____ diagram of PV power plant shows various losses at different stage of conversion
a) Sankey
b) Fish
c) Bone
d) Mesh

206. Inverter AC power output is 95 KW and 5 % of inverter power loss. The inverter DC power input is _____
a) 100
b) 120
c) 150
d) 200

207. Inverter DC power input is 90 KW and 10 % of cable loss. The PV array DC power output is _____
a) 100
b) 120
c) 150
d) 200

208. A solar plant has generated 200 units per year and 180 units of electricity fed to the grid by considering all the losses. The performance ratio is _____
   a) 80 %
   b) 70 %
   c) 50 %
   d) 40 %

209. Module surface temperature is measured by _____
   a) Thermocouple
   b) Ammeter
   c) Anemometer
   d) None

210. If polarity of PV module is reversed, it results in damage of _____
   a) Bypass Diode
   b) Inverter Input
   c) Option a & b
   d) None

211. If Module Voc 1 is 12 V, Isc 1 is 2 A, Module 2 Voc 2 is 12 V, Isc 2 is 2 A, Module 3 Voc 3 is 12 V, Isc 3 is 2 A.
The total open circuit voltage Voc and short circuit current Isc is _____, in parallel connection.
   a) 12 V, 2 A
   b) 12 V, 6 A
   c) 24 V, 6 A
   d) 36 V, 4 A

212. Range of specific gravity is _____
   a) 1.1 to 1.4
   b) 1.5 to 2
   c) 2 to 2.5
   d) 3 to 3.5

213. Multimeter is used to do the _____ test
   a) Continuity
   b) Polarity
   c) Option a & b
   d) None

214. SoC has decreased if its specific gravity is _____
   a) Decreases
   b) Increases
   c) Option a & b
   d) None

215. One simple test for cable check is _____
    a) Continuity Test
    b) Clamp Test
    c) Clip Test
    d) None

216. Testing of cable is done by _____
    a) Continuity Test
    b) HV Test
    c) IR Test
    d) All of these

217. Cable failure is due to _____
    a) Low IR Value
    b) Over Voltage
    c) Heating
    d) All of these

218. The component to be tested before installing the PV system is _____
    a) PV Module
    b) Inverter
    c) Monitoring System
    d) All of these

219. _____ drawing to be checked before installing of PV system
    a) Electrical Layout
    b) Mechanical Layout
    c) Civil Layout
    d) All of these

220. Type of test for PV system is _____
    a) Open Voltage Test
    b) Short Circuit Test
    c) Polarity Test
    d) All of these

221. Type of test done for battery system is _____
    a) Polarity Test
    b) Capacity Test
    c) Hydrometer Test
    d) All of these

222. Quality Control is known as _____
   a) Identifying defects after production of products
   b) Identifying defects before production of products
   c) Option a & b
   d) None

223. Quality Assurance is known as _____
   a) Prevent defects with a focus on the process used to make the product
   b) Defects find out after delivery to customer end
   c) Option a & b
   d) None

224. QA means _____
   a) Quality Assurance
   b) Quantity Assured
   c) Quantity Acquisition
   d) None

225. QC means _____
   a) Quality Control
   b) Quantity Condition
   c) Quality Concern
   d) None

226. DAS is helpful to _____ the complete PV system
   a) Record
   b) Monitor
   c) Option a & b
   d) None

227. If DAS shows wrong reading, we should _____
   a) Replace with new system
   b) Repair it
   c) Redo the calibration
   d) All of these

228. PV Test reports includes _____
   a) Visual inspection record and observations
   b) Identification of circuits tested, tests performed, and record of measurements
   c) Interpretation and summary of results
   d) All of these

229. _____ is the method of testing in PV plant
    a) Continuity Testing   b) Polarity Testing
    c) Voltage and Current Testing   d) All of these

230. _____ instrument is used to measure the solar radiation
    a) Pyranometer   b) Megger
    c) Multimeter   d) None

231. _____ instrument is used to measure the insulation resistance value
    a) Pyranometer   b) Megger
    c) Multimeter   d) None

232. _____ test is used to avoid short circuit
    a) Polarity   b) Power
    c) Pole   d) None

233. Module to Module interconnection is verified by the _____
    a) Polarity Test   b) Wattmeter Test
    c) Option a & b   d) None

234. Performance test of PV plant includes _____
    a) Thermal Imaging   b) Power Quality Analysis
    c) Shading Analysis   d) All of these

235. Performance test of PV plant includes _____
    a) Inverter Efficiency Tests
    b) Maximum Power Point Tracking Tests
    c) Voltage & Current testing
    d) All of these

236. If PV output is 5 V, 1 A _____ connection gives the output of 1 Amps & 20 Volts
    a) Series
    b) Parallel
    c) Parallel - Series
    d) Series - Parallel

237. If PV output of 5 V, 1 A _____ number of module is required to get the output of 1 Amps & 20 Volts
    a) 1
    b) 2
    c) 4
    d) 5

238. If PV output of 5 V, 1 A _____ connection gives the output of 5 Amps & 5 Volts
    a) Series
    b) Parallel
    c) Parallel - Series
    d) Series - Parallel

239. If PV output of 5 V, 1 A _____ number of module is required to get the output of 5 Amps & 5 Volts
    a) 1
    b) 2
    c) 5
    d) 10

240. If 1000 W fan is used for 6 hours. The average cost of electricity is Rs. 2 per kWh. The cost of electricity consumed by the fans is _____
    a) INR 12
    b) INR 9
    c) INR 6
    d) INR 3

241. The electricity bill is _____, if consuming the monthly bill of 100 units and cost of electricity is ₹ 3 per kWh
    a) 300
    b) 30
    c) 3
    d) 0.30

242. _____ parameter is matched with grid connection during inverter synchronization
   a) Frequency
   b) Voltage
   c) Option a & b
   d) None

243. _____ will happen during grid synchronization with inverter
   a) Voltage Fluctuation
   b) Frequency Variation
   c) Phase Unbalance
   d) All of these

244. The anti - islanding system is operates when _____
   a) Disconnect from Grid
   b) More than 5 % Voltage Fluctuation
   c) Frequency is not in the range of 47.5 to 50.5
   d) All the above

245. The PV Plant is tested to find the _____
   a) Performance Ratio Assessment
   b) Diagnosis of Malfunctions
   c) Performance Monitoring
   d) All of these

Chapter 8

Maintain Solar PV System

1. The distance between two earth pits should be _____ meter
   a) 1
   b) 2
   c) 3
   d) 4

2. Life of the battery is around _____ years
   a) 2 to 5
   b) 7 to 8
   c) 9 to 10
   d) 20 to 25

3. The factors affecting the maintenance of PV plant is due to the _____
   a) Installer where does not follow actual design of the system
   b) Labelling is incorrect
   c) Loose cables
   d) All of these

4. The factors affecting the maintenance of PV plant is due to the _____
   a) Poor Wiring
   b) Labelling not present
   c) No Earthing
   d) All of these

5. The factors affecting the maintenance of PV plant is due to the _____
   a) Lightning Protection
   b) Rust in solar structure
   c) Badly placed Sensors
   d) All of these

6. Mounting structure is blows up if _____
   a) Roof perforation without adequate sealing methods
   b) Actual wind speed is more than designed value
   c) Option a & b
   d) None

7. The solar module fails if _____
   a) Short circuited when wrong connection
   b) No ventilation
   c) Over heats due to soil deposit
   d) All of these

8. PV modules starts to corrode when _____
   a) Located near Salt Water
   b) No Galvanisation
   c) Option a & b
   d) None

9. The structure failure occurs due to _____
   a) Inadequate Mounting
   b) Structure not mounted on Concrete Bases
   c) Option a & b
   d) None

10. The PV plant get tripped repeatedly due to the _____
    a) Inappropriate Inverter
    b) Undersized Cables
    c) Undersized Power Optimizer
    d) All of these

11. Structure failure happens while installing the panel due to the _____
    a) Age Factor
    b) Over Load of Roof
    c) Option a & b
    d) None

12. Corrosion of structure is due to the _____
    a) No UV Resistant
    b) No Sunlight Resistant
    c) Option a & b
    d) None

13. Inverter failure happens when PV is _____
    a) Placed directly exposed to the Sunlight
    b) Insufficient Ventilation
    c) Improper cable design from the PV array to JB box
    d) All of these

14. Cable failure is due to the _____
    a) Tight / Loose cables
    b) Improper cable support with exposure to physical damage
    c) Multiple cables entering a single conductor cable gland
    d) All of these

15. Battery failure is due to the _____
    a) Exposed of Sunlight Directly
    b) Exposed of High Temperature
    c) Installation closer to Flammable Materials
    d) All of these

16. The life of battery is higher, if we maintained _____
    a) Its surface by Cleaning Regularly
    b) Its Electrolyte Level is maintained constantly
    c) Option a & b
    d) None

17. The desired earthing value in earth pit is _____
    a) Less than 5 Ohm
    b) More than 10 Ohm
    c) Option a & b
    d) None

18. _____ is essential to achieve the best performance in PV plant
    a) Routine Cleaning
    b) Scheduled Maintenance
    c) Proper Monitoring System
    d) All of these

19. The important key factor of a PV plant is _____
    a) Good Design   b) Proper O&M
    c) Option a & b   d) None

20. Battery cycle ranges from _____
    a) 500 to 1500   b) 2000 to 3000
    c) 3000 to 4000   d) 5000 to 6000

21. If battery is 100 % fully charged, then the Depth of Discharge (DoD) of battery is _____
    a) 0 %   b) 100 %
    c) 90 %   d) 80 %

22. If I say the battery have delivered 30 % of its energy and remaining of 70 % energy reserved. The DoD of battery is _____
    a) 30 %   b) 70 %
    c) 80 %   d) 90 %

23. Life of the battery is less, if the daily Discharge Level is _____
    a) 20 %   b) 30 %
    c) 40 %   d) 80 %

24. Life of the battery is high, if the daily Discharge Level is _____
    a) 40 %   d) 70 %
    c) 80 %   b) 90 %

25. Life of battery is _____ years, if DoD of battery is 40 % daily
    a) 0.5 to 1   b) 1.5 to 2
    c) 3 to 5   d) None

26. Life of battery is _____ years, if DoD of battery is 80 % daily
    a) 1   b) 9
    c) 10   d) None

27. Battery is automatically _____ if not utilized for two months
    a) Charged   b) Discharged
    c) Option a & b   d) None

28. At higher temperature batteries can deliver _____ current than lower temperature
    a) Lower   b) Higher
    c) Option a & b   d) None

29. _____ is required for liquid vented lead acid batteries
    a) Refilling of Water   b) No Refilling of Water
    c) Option a & b   d) None

30. _____ is required for sealed lead acid batteries
    a) No Refilling of Water   b) Refilling of Water
    c) Option a & b   d) None

31. The battery capacity is "C". The C-Rating of the battery for 2 hours of charging is _____
    a) C / 2   b) C / 0.5
    c) 2 /C   d) 0.5 C

32. The battery capacity is "C". The C-Rating of the battery for 20 hours of charging is _____
    a) C / 2   b) C / 0.05
    c) C 2   d) 0.5 C

33. The battery capacity is 50 Ah. The C-Rating of the battery for 5 hours of charging is _____
    a) C / 5   b) C / 0.05
    c) C 2   d) 0.5 C

34. The battery capacity is 80 Ah. The C-Rating of the battery for 10 hours of charging is _____
    a) C / 10
    b) C / 0.05
    c) C 5
    d) 0.5 C

35. The battery capacity is 200 Ah. The C-Rating of the battery for 20 hours of charging is _____
    a) C / 20
    b) C / 0.05
    c) C 5
    d) 0.5 C

36. _____ type of electrode is used for solar system
    a) Tubular
    b) Flat
    c) Option a & b
    d) None

37. SMF battery is _____
    a) Maintenance Free
    b) Advanced
    c) Old Technology
    d) None

38. _____ is the gap between two panels placed in row for maintenance access
    a) 0.5 to 1 meter
    b) 1.5 to 2 meter
    c) 3.5 to 4 meter
    d) 5 to 8 meter

39. Resistance value of PV array mounting structure to ground is _____
    a) 0.1 ohm
    b) 5 ohm
    c) 10 ohm
    d) 20 ohm

40. _____ switch is isolated between PV array to Inverter
    a) DC Disconnect
    b) AC Disconnect
    c) Option a & b
    d) None

41. _____ should be checked before charging the battery
    a) Cleanliness of Battery, Rack / Cabinet, Battery Room
    b) Inspect for Cracks, Leaks, any corrosion
    c) Check Connection Torques
    d) All of these

42. _____ to be measured during battery testing
    a) Connection Resistances b) Voltage and Charging Current
    c) Specific Gravity   d) All of these

43. _____ should be checked before installing earthing pits system
    a) Size of the Conductor   b) Available Area
    c) Chemical Materials   d) All of these

44. After installing the inverter, connect with _____
    a) Battery   b) PV Modules
    c) Both a & b   d) None

45. After installing the earthing pits, _____ test should be done
    a) Megger   b) Level
    c) Clean   d) Depth

46. The earthing value is best at _____
    a) < 5 ohm   b) 10 ohm
    c) 15 ohm   d) 20 ohm

47. Earthing is done to find the _____
    a) Fault current flow through earth
    b) Rated current flow through earth
    c) Half of the rated current flow through earth
    d) None

48. _____ volt is between earth and ground
    a) < 2
    b) 5
    c) 8
    d) 10

49. _____ should be kept at office after installation of PV system
    a) Technical Manuals
    b) Safety Manuals
    c) Option a & b
    d) None

50. Grounding is required for _____
    a) PV Module
    b) Inverter
    c) Structure
    d) All of these

51. _____ operates the anti - islanding protection mode
    a) Voltage fluctuation limit of +/ – 5 %
    b) Not meet the frequency range of 47.5 to 50.5 HZ
    c) Disconnect from the Grid
    d) All of these

52. The minimum acceptable value of insulation resistance is _____
    a) Mega ohm
    b) Kilo ohm
    c) ohm
    d) None

53. _____ is the best material for earthing
    a) Copper
    b) Aluminum
    c) GI
    d) Iron

54. _____ is a conducting material
    a) Copper
    b) GI
    c) Aluminum
    d) All of these

55. _____ is excellent conductor than GI
    a) Copper
    b) Glass
    c) Rubber
    d) Wood

56. Type of earthing used is _____
    a) Rod   b) Pipe
    c) Plate   d) All of these

57. Earthing is used to protect from an electric shock is _____
    a) Earthing   b) Welding
    c) Cutting   d) None

58. Main purpose of earthing is _____
    a) Fault current flow to earth   b) To ensure the equipment safety
    c) Option a & b   d) None

59. Lighting arrestor is to be fixed against _____
    a) Lighting   b) Birds
    c) Animals   d) None

60. Lighting arrestor is fixed at the _____ of the building
    a) Top   b) Bottom
    c) Left Side   d) Right Side

61. Battery area should be an _____
    a) Ventilated Room   b) Low Temperature Room
    c) Option a & b   d) None

62. Minimum distance between two fixed modules should be _____
    a) 5 mm   b) 15 mm
    c) 20 mm   d) 25 mm

63. The purpose of grounding is _____
    a) Human Safety   b) Equipment Safety
    c) Option a & b   d) None

64. The battery bank is located at _____
    a) Ventilated Room   b) Clean Room
    c) Option a & b   d) None

65. _____ is the cause for degradation of module
    a) Aging   b) Cleanness
    c) Earthing   d) All of these

66. The expected life of solar plant is _____ years
    a) 1 to 2   b) 15 to 25
    c) 5 to 8   d) 2 to 5

Chapter 9

Solar PV Project Lifecycle

1. The life cycle of a solar panel is _____ years
   a) 20 to 25
   b) 30 to 35
   c) 40 to 55
   d) None

2. LCA means _____
   a) Life Cycle Assessment
   b) Life Cycle Assignment
   c) Option a & b
   d) None

3. Environmental study of LCA includes of _____
   a) Raw Materials, Manufacturing Process
   b) Distribution and Maintenance Process
   c) Recycling and Final Disposal Process
   d) All of these

4. LCA study has the following issues _____
   a) Resource Consumption
   b) Health of Human Being
   c) Environment Effects
   d) All of these

5. LCC means _____
   a) Life Cycle Cost
   b) Life Cycle Cause
   c) Option a & b
   d) None

6. The cost included in LCC study is _____
   a) Raw Materials Cost
   b) Manufacturing and Installation Cost
   c) Disposal and Recycling
   d) All of these

7. _____ deals with environmental impact study of solar panel
   a) Life Cycle Assessment
   b) Life Cycle Cost
   c) Option a & b
   d) None

8. _____ deals with cost related study of solar panel
   a) Life Cycle Cost
   b) Life Cycle Assessment
   c) Option a & b
   d) None

Chapter 10

Determine the Financial Viability of Solar PV Power Plant

1. _____ is important for the financial part of PV system
   a) Pay Back Period
   c) ROI
   b) IRR
   d) All of these

2. ROI means _____
   a) Return on Investment
   c) Option a & b
   b) Risk on Investment
   d) None

3. IRR means _____
   a) Internal Rate of Return
   c) Investment Rate of Return
   b) Initial Rate of Return
   a) None

4. The payback period is a _____
   a) Investment / Saving
   c) Saving / Investment
   b) Investment x Saving
   d) None

5. A small house installing of 1 KWp solar plant at ₹ 60,000 as initial investment, expected at ₹ 15000 as energy saving per year. The simple payback period in years is _____
   Formula: Payback Period = Investment / (Saving – Maintenance)
   a) 4
   c) 6
   b) 5
   d) 3

6. A house installing of 1 KWp solar plant at ₹ 60,000 as initial investment, expected at ₹ 15,000 as energy saving per year, and maintenance cost at ₹ 5,000 year is added as extra. The payback period is _____ years
   Formula : Payback Period = Investment / (Saving - Maintenance)
   a) 6 b) 5
   c) 4 d) 3

7. The payback period of solar plant is _____ years
   a) 4 to 7 b) 7 to 8
   c) 8 to 9 d) 9 to 10

Chapter 11

Maintain Personal Health & Safety at Project Site

1. PPE means _____
   a) Personal Protective Equipment
   b) Personal Protection Equipment
   c) Permitted Protective Equipment
   d) Personal Protective Equation

2. The regularly inspected equipment is _____
   a) Monitoring Equipment   b) Safety Equipment
   c) Option a & b   d) None

3. _____, it will cause the death
   a) The installer walks on the panels while live
   b) The inverter is installed while live
   c) Doing the electrical work while live
   d) All of these

4. The structure repair should not be done when there is _____
   a) Strong Wind Situation   b) Heavy Rain Falls
   c) Dust Storm   d) All of these

5. You must have the equipment _____ before going to do electrical safety audit
   a) Manual   b) Warranty
   c) Test Certificate   d) All of these

6. The safety equipment used while doing work at height is _____
   a) Belt
   b) Helmet
   c) Shoe
   d) All of these

7. The safety equipment used while doing work at substation is _____
   a) Cloves
   b) Helmet
   c) Shoe
   d) All of these

8. The safety equipment used while doing work at heavy industries is _____
   a) Nose mask
   b) Ear plug
   c) Shoe
   d) All of these

9. The safety equipment used while doing welding work at site is _____
   a) Safety cloth
   b) Goggle
   c) Shoe
   d) All of these

10. OHS means _____
    a) Occupational Health and Safety
    b) One Hour Security
    c) Our Home Secret
    d) None

11. _____ needs safety training
    a) Supervisor
    b) Employee
    c) Manager
    d) All of these

12. _____ is needed while doing electrical work
    a) Safety of Human by wearing PPE
    b) Grounding of Panel
    c) Disconnect from Power Supply
    d) All of these

13. _____ is utmost importance at the industrial area
    a) Workplace Safety   b) Net Worth of Factory
    c) Factory EB Bill   d) None

14. You must do _____ before doing any electrical maintenance on the HT/LT panel
    a) Log - Out / Tag - Out   b) Check In
    c) Exit Out   d) Exit In

15. Hazardous Area is identified by _____
    a) Danger Sign   b) Explosive Symbol
    c) Siren Alarm   d) All of these

16. The usage of correct PPE is done by _____
    a) Guidance from Safety Manager
    b) Reading the Working Manual of PPE
    c) Practicing PPE while attending Training Session
    d) All of these

17. The first aid box is available at _____
    a) Hospital   b) Fire Station
    c) Industry   d) All of these

18. First aid box contains _____
    a) Splint & Antiseptic Wipes   b) Gauze Pads
    c) Adhesive Bandage   d) All of these

19. _____ is a part of PPE
    a) Shoes   b) Helmet
    c) Gloves   d) All of these

20. _____ is a part of 5S
    a) Sort
    c) Standardize
    b) Shine
    d) All of these

21. _____ is not a part of 5S
    a) Size
    c) Sustain
    b) Set In Order
    d) None

22. SOP means _____
    a) Standard Operating Procedure
    b) Standard Operating Protocol
    c) Standard Operating Process
    d) None

23. _____ is the safety tool kit
    a) Safety Helmet
    c) Safety Goggles
    b) Safety Hand - Gloves
    d) All of these

24. _____ is the safety tool kit
    a) Safety Harness
    c) Option a & b
    b) Reflective Jacket
    d) None

Answer Sheet

Chapter 1				
1. A	2. B	3. A	4. B	5. D
6. A	7. A	8. A	9. A	10. C
11. A	12. D	13. C	14. A	15. C
16. D	17. D	18. D	19. B	20. D
21. A	22. D	23. D	24. D	25. D
26. D	27. A	28. B	29. C	30. D
31. D	32. C	33. A	34. A	35. A
36. A	37. A	38. A	39. A	40. C
41. D	42. A	43. D	44. D	45. D
46. A	47. D	48. C	49. C	50. A
51. C	52. A	53. A	54. C	55. A
56. A	57. A	58. A	59. C	60. A
61. A	62. A	63. A	64. B	65. C
66. B	67. B	68. C	69. B	70. C
71. A	72. B	73. D		

Chapter 2				
1. A	2. A	3. A	4. A	5. A
6. A	7. A	8. A	9. A	10. D
11. D	12. A	13. A	14. A	15. A
16. A	17. A	18. A	19. A	20. A
21. A	22. D	23. A	24. D	25. A

142 | Answer Sheet

26. C	27. A	28. D	29. D	30. A
31. D	32. A	33. D	34. A	35. A
36. A	37. A	38. A	39. A	40. A
41. A	42. A	43. D	44. B	45. A
46. D	47. B	48. A	49. A	50. B
51. A	52. A	53. A	54. A	55. C
56. D	57. D	58. D	59. D	60. A
61. A	62. A	63. A	64. A	65. A
66. A	67. B	68. A	69. B	70. D
71. A	72. A	73. A	74. A	75. A
76. A	77. A	78. A	79. A	80. D
81. A	82. A	83. A	84. A	85. A
86. A	87. A	88. A	89. C	90. A
91. A	92. B	93. A	94. B	95. A
96. D	97. A	98. A	99. A	100. A
101. D	102. D	103. D	104. D	105. A
106. A	107. D	108. A	109. A	110. A
111. A	112. C	113. A	114. A	115. A
116. D	117. A	118. C	119. A	120. C
121. C	122. D			

Chapter 3				
1. A	2. A	3. A	4. A	5. A
6. A	7. C	8. C	9. A	10. D
11. C	12. A	13. A	14. A	15. A
16. A	17. A	18. A	19. A	20. D
21. A	22. A	23. D	24. A	25. A
26. A	27. A	28. A	29. A	30. D
31. A	32. D	33. C	34. A	35. A
36. A	37. A	38. A	39. A	40. A
41. A	42. A	43. D	44. A	45. A
46. A	47. A	48. A	49. A	50. A
51. D	52. A	53. A	54. A	55. A
56. C	57. A	58. A	59. A	60. A

Answer Sheet | 143

61. D	62. D	63. D	64. D	65. D
66. D	67. A	68. A	69. A	70. D
71. D	72. D	73. D	74. D	75. C
76. A	77. D	78. A	79. A	80. D
81. D	82. D	83. D	84. D	85. A
86. C	87. A	88. A	89. D	90. A
91. A				

Chapter 4				
1. D	2. D	3. D	4. D	5. D
6. D	7. D	8. D	9. C	10. C
11. C	12. C			

Chapter 5				
1. A	2. A	3. A	4. C	5. A
6. A	7. A	8. C	9. A	10. A
11. A	12. A	13. A	14. C	15. D
16. A	17. A	18. A	19. C	20. A
21. D	22. A	23. A	24. A	25. A
26. A	27. A	28. A	29. D	30. D
31. A	32. A	33. A	34. A	35. A
36. A	37. A	38. A	39. A	40. A
41. A	42. C	43. A	44. A	45. A
46. A	47. A	48. A	49. C	50. A
51. A	52. A	53. A	54. B	55. A
56. A	57. A	58. A	59. A	60. A
61. A	62. A	63. A	64. A	65. A
66. A	67. A	68. A	69. A	70. A
71. A	72. A	73. A	74. A	75. A
76. A	77. E	78. A	79. A	80. A
81. D	82. A	83. A	84. A	85. A
86. A	87. A	88. A	89. A	90. B
91. A	92. A	93. C	94. A	95. D

96. D	97. B	98. A	99. A	100. D
101. A	102. A			.

Chapter 6				
1. C	2. D	3. D	4. D	5. D
6. D	7. D	8. A	9. D	10. B
11. A	12. C	13. B	14. D	15. B
16. C	17. C	18. A	19. A	20. D
21. A	22. A	23. A	24. B	25. C
26. B	27. C	28. A	29. A	30. A
31. B	32. B	33. A	34. A	35. B
36. A	37. A	38. A	39. A	40. A
41. B	42. D	43. A	44. D	45. A
46. B	47. A	48. A	49. A	50. A
51. A	52. A	53. C	54. C	55. D
56. C	57. D	58. B	59. C	60. D
61. C	62. C	63. A	64. A	65. A
66. A	67. A	68. A	69. B	70. C
71. B	72. A	73. B	74. A	75. C
76. B	77. B	78. B	79. C	80. A
81. C	82. D			

Chapter 7				
1. A	2. A	3. D	4. A	5. A
6. A	7. A	8. A	9. D	10. A
11. A	12. C	13. D	14. C	15. C
16. A	17. A	18. A	19. A	20. A
21. C	22. B	23. C	24. A	25. A
26. A	27. A	28. B	29. B	30. B
31. C	32. C	33. C	34. A	35. A
36. B	37. A	38. A	39. D	40. D
41. D	42. A	43. A	44. A	45. A

46. A	47. A	48. A	49. D	50. A
51. A	52. A	53. A	54. A	55. C
56. A	57. C	58. C	59. A	60. A
61. D	62. A	63. A	64. A	65. A
66. D	67. D	68. A	69. A	70. B
71. C	72. B	73. C	74. A	75. C
76. B	77. C	78. B	79. B	80. A
81. A	82. A	83. A	84. A	85. A
86. A	87. A	88. D	89. B	90. C
91. D	92. B	93. D	94. A	95. C
96. B	97. A	98. D	99. A	100. A
101. C	102. A	103. C	104. A	105. A
106. D	107. C	108. A	109. B	110. A
111. A	112. A	113. B	114. A	115. A
116. A	117. A	118. A	119. B	120. B
121. A	122. A	123. A	124. B	125. A
126. B	127. B	128. C	129. D	130. C
131. D	132. A	133. A	134. A	135. C
136. D	137. C	138. C	139. D	140. D
141. C	142. D	143. B	144. C	145. C
146. A	147. A	148. A	149. C	150. A
151. C	152. A	153. D	154. A	155. A
156. A	157. A	158. B	159. A	160. D
161. A	162. A	163. A	164. A	165. A
166. A	167. A	168. A	169. A	170. A
171. A	172. A	173. A	174. A	175. A
176. C	177. A	178. D	179. A	180. C
181. C	182. B	183. D	184. B	185. A
186. A	187. A	188. B	189. C	190. A
191. B	192. C	193. A	194. A	195. B
196. B	197. A	198. B	199. A	200. A
201. A	202. B	203. B	204. A	205. A
206. A	207. A	208. A	209. A	210. A
211. B	212. A	213. A	214. A	215. A

216. D	217. D	218. D	219. D	220. D
221. D	222. A	223. A	224. A	225. A
226. C	227. D	228. D	229. D	230. A
231. B	232. A	233. A	234. D	235. D
236. A	237. C	238. B	239. C	240. A
241. A	242. C	243. D	244. D	245. D

Chapter 8				
1. B	2. A	3. D	4. D	5. D
6. C	7. D	8. C	9. C	10. D
11. C	12. C	13. D	14. D	15. D
16. C	17. A	18. D	19. C	20. A
21. A	22. A	23. D	24. A	25. C
26. A	27. B	28. B	29. A	30. A
31. A	32. A	33. A	34. A	35. A
36. C	37. A	38. A	39. A	40. C
41. D	42. D	43. D	44. C	45. A
46. A	47. A	48. A	49. C	50. D
51. D	52. A	53. A	54. D	55. A
56. D	57. A	58. A	59. A	60. A
61. C	62. A	63. C	64. C	65. D
66. B				

Chapter 9				
1. A	2. A	3. D	4. D	5. A
6. D	7. A	8. A		

Chapter 10				
1. D	2. A	3. A	4. A	5. A
6. A	7. A			

Chapter 11				
1. A	2. C	3. D	4. D	5. D
6. D	7. D	8. D	9. D	10. A
11. D	12. C	13. A	14. A	15. D
16. D	17. D	18. D	19. D	20. D
21. A	22. A	23. D	24. C	

Bibliography

Asian Development Bank (ADB)
https://www.adb.org/

Asia Pacific Energy Research Centre (APERC)
https://aperc.ieej.or.jp/

Alliance for Rural Electrification (ARE)
https://www.ruralelec.org/

American Council on Renewable Energy (ACORE)
https://acore.org/

Associação Portuguesa de Energias Renováveis (APREN)
https://www.apren.pt/

Association for Renewable Energy of Lusophone Countries (ALER)
www.aler-renovaveis.org/en/home/

Agency for Non-conventional Energy and Rural Technology (ANERT)
https://www.anert.gov.in/

Arunachal Pradesh Energy Development Agency (APEDA)
http://www.apeda.nic.in/

Assam Energy Development Agency
http://assamrenewable.org/

Bihar Renewable Energy Development Agency
www.breda.bih.nic.in/

Chinese Renewable Energy Industries Association (CREIA)
www.creia.net/creiaen.html

Clean Energy Council (CEC)
https://www.cleanenergycouncil.org.au/

Climate Action Network International (CAN-I)
www.climatenetwork.org/

Council on Energy, Environment and Water (CEEW)
https://www.ceew.in/

Chhattisgarh State Renewable Energy Development Agency (CREDA)
www.creda.in/

ECOWAS Centre for Renewable Energy and Energy Efficiency (ECREEE)
www.ecreee.org/

European Commission (EC)
https://ec.europa.eu/commission/index_en

Electricity sector in India
https://en.wikipedia.org/wiki/Electricity_sector_in_India

European Renewable Energies Federation (EREF)
www.eref-europe.org/

Global Off-Grid Lighting Association (GOGLA)
https://www.gogla.org/the-voice-of-the-off-grid-solar-energy-industry

Global Solar Council (GSC)
www.globalsolarcouncil.org/

Global Wind Energy Council (GWEC)
https://www.gwec.net/

Goa Energy Development Agency
http://geda.goa.gov.in

Gujarat Energy Development Agency (GEDA)
https://geda.gujarat.gov.in/

Global Alliance for Clean Cookstoves (GACC)
https://cleancookingalliance.org/

Global Forum on Sustainable Energy (GFSE)
https://www.gfse.at/

Greenpeace International
https://www.greenpeace.org/international/

Global Environment Facility (GEF)
https://www.thegef.org/

Haryana Renewal Energy Development Agency (HAREDA)
https://hareda.gov.in/

HIMURJA - Himachal Pradesh Energy Development Agency
http://himurja.hp.gov.in/

International Institute for Applied Systems Analysis (IIASA)
http://www.iiasa.ac.at/

International Solar Energy Society (ISES)
https://www.ises.org/

Integrated Rural Energy Programme (IREP)
http://web.iitd.ac.in

Indian Renewable Energy Federation (IREF)
https://iref.net.in/

International Geothermal Association (IGA)
https://www.geothermal-energy.org/

International Hydropower Association (IHA)
https://www.hydropower.org/

International Energy Agency
https://www.iea.org/topics/renewables/solar/

International Energy Agency (IEA)
https://www.iea.org/

International Solar Alliance
http://isolaralliance.org/

152 | Bibliography

ICLEI – Local Governments for Sustainability, South Asia
southasia.iclei.org/

International Electrotechnical Commission (IEC)
https://www.iec.ch

Institute for Sustainable Energy Policies (ISEP)
https://www.isep.or.jp/en

Indian Renewable Energy Development Agency Ltd.
https://ireda.in/

Indian Renewable Energy Development Agency Limited (IREDA)
https://ireda.in/

International Institute of Sustainable Development (IISD)
https://www.iisd.org/

imec R&D, nano electronics and digital technologies
https://www.imec-int.com

Jammu & Kashmir Energy Development Agency (JAKEDA)
http://jakeda.jk.gov.in/

Jharkhand Renewable Energy Development Agency
www.jreda.com/

Karnataka Renewable Energy Development Agency Ltd.
https://kredlinfo.in/

MP Urja Vikas Nigam Ltd.
www.mprenewable.nic.in/

Maharashtra Energy Development Agency (MEDA)
https://www.mahaurja.com/

Manipur Renewable Energy Development Agency (MANIREDA)
http://manireda.com/

Meghalaya Non-conventional & Rural Energy Development Agency
http://mnreda.gov.in/

Mercom magazine
https://mercomindia.com

Ministry of Power
https://powermin.nic.in/

Mali Folke Center (MFC)
https://www.malifolkecenter.org/

National Institute of Solar Energy
https://nise.res.in/

National Institute of Wind Energy
https://niwe.res.in

National Institute of Solar Energy
www.nise.res.in/

Nagaland Renewable Energy Development Agency
http://nre.nagaland.gov.in/

Non-conventional Energy Development Agency (NEDA)
http://www.upneda.org.in/

Non-Conventional Energy Development Corporation of Andhra Pradesh (NEDCAP) Ltd.
http://nredcap.in/

National Renewable Energy Laboratory (NREL)
https://www.nrel.gov/

National Offshore Wind Energy Authority
https://www.windpoweroffshore.com

Odisha Renewable Energy Development Agency (OREDA)
http://oredaodisha.com

PV magazine
https://www.pv-magazine.com

Punjab Energy Development Agency
http://www.peda.gov.in/

Photovoltaic Solar Testing Specifications
http://www.cszindustrial.com/Products/Custom-Designed-Chambers/Solar-Panel-Testing-Chamber/Solar-Testing-Specifications.aspx

Partnership for Sustainable Low Carbon Transport (SLoCaT)
http://www.slocat.net/

Rajasthan Renewable Energy Corporation Limited
http://energy.rajasthan.gov.in/content/raj/energy-department/rrecl/en/home.html

Renewable Energy Agency Puducherry (REAP)
Rooftop Solar Grid Engineer
http://sscgj.in/wp-content/uploads/2016/06/SGJ_Q0106_Rooftop-Solar-Grid-Engineer-Model-Curriculum.pdf

Rooftop Solar Photovoltaic Entrepreneur
http://sscgj.in/wp-content/uploads/2017/06/SGJ_Q0104_Rooftop-Solar-Photovoltaic-Entrepreneur-Model-Curriculum.docx-1.pdf

REN21: Home
www.ren21.net

Renewable Energy Institute (REI)
https://www.renewable-ei.org/en/

Renewable Energy Solutions for the Mediterranean (RES4MED)
https://www.res4med.org/

Regional Center for Renewable Energy and Energy Efficiency (RCREEE)
www.rcreee.org/

Sikkim Renewable Energy Development Agency
http://www.sreda.gov.in

Skill Council for Green Jobs
http://sscgj.in/

Solar Energy Corporation of India Limited
http://seci.co.in/

Solar Power Europe
http://www.solarpowereurope.org/wp-content/uploads/2018/09/Global-Market-Outlook-2018-2022.pdf

Solar Energy Events
https://10times.com/top100/solar-energy

Solar PV Installer - Civil
http://sscgj.in/wp-content/uploads/2017/06/Model-Curricullum_SGJ-Q0103_Solar-PV-Installer-Civil.pdf

Solar PV Installer - Suryamitra
http://sscgj.in/wp-content/uploads/2016/06/SGJ-Q0101_Solar-PV-Installer-Suryamitra_Model-Curricullum.pdf

Solar PV Installer - Electrical
http://sscgj.in/wp-content/uploads/2017/06/Model-Curricullum_SGJ-Q0102_Solar-PV-Installer-Electrical.pdf

Solar Proposal Evaluation Specialist
http://sscgj.in/wp-content/uploads/2017/06/SGJ_Q0105_-Solar-proposal-Evaluation-Specialist-Model-Curriculum-1.pdf

Solar Book
Fundamentals, Technologies And Applications – Chetan Singh Solanki. PHI Learning Pvt. Ltd.

South African National Energy Development Institute (SANEDI)
https://www.sanedi.org.za

Tamil Nadu Energy Development Agency (TEDA)
http://teda.in/

Telangana New and Renewable Energy Development Corporation Limited [TNREDCL]
http://tsredco.telangana.gov.in/

Tripura Renewable Energy Development Agency
http://treda.nic.in/

TERI: Innovative Solutions for Sustainable Development - India
https://www.teriin.org/

Uttarakhand Renewable Energy Development Agency (UREDA)
http://ureda.uk.gov.in

United Nations Development Programme (UNDP)
https://www.undp.org/content/undp/en/home.html

United Nations Environment Programme (UN Environment)
www.unenvironment.org/

United Nations Industrial Development Organization (UNIDO)
https://www.unido.org/

West Bengal Renewable Energy Development Agency (WBREDA)
http://www.wbreda.org/

World energy consumption data
https://en.wikipedia.org/wiki/World_energy_consumption

World Bioenergy Association (WBA)
https://worldbioenergy.org

World Wind Energy Association (WWEA)
https://wwindea.org/

World Bank (WB)
https://www.worldbank.org/

WFES Abu Dhabi - Powering the Future of Energy and Sustainability
https://www.worldfutureenergysummit.com/

World Council for Renewable Energy (WCRE)
https://www.wcre.de/

World Future Council (WFC)
https://www.worldfuturecouncil.org/

World Resources Institute (WRI)
https://www.wri.org/

Zoram Energy Development Agency (ZEDA)
https://zeda.mizoram.gov.in/

Index

A

AC Circuit, 23, 52, 95
ACB, xvii, 43
Access of Roof, 2, 28
Acre, 8, 30
Actual Capacity, 78
Actual Load, 97
Adhesive Bandage, 139
Advanced, 128
Advantage, 21-22
Age Factor, 124
Air Mass, 33
Air Speed, 43
Aluminium, 44, 67
Aluminium Wire, 67
Ammeter, 98-99, 109, 116
Amount of Light Falls, 6
Ampere Hour, 62, 88-89, 91
Anemometer, 34, 41, 43, 116
Angle of Light Falls, 6
Antiseptic Wipes, 139
Avoid Hot Spots, 87
Azimuth, 2

B

Baking Soda, 48
Balance of System, xvii, 43
Batteries, 37-38, 73-76, 78-79, 91, 112, 127
Battery, xxiii-xxv, 3, 5, 24, 26-27, 36-38, 40, 42-44, 48, 52, 62-63, 66, 71-76, 78-79, 88-91, 105-112, 117, 123, 125-129, 131-132
Battery Cycle Range, 126
Battery Efficiency, xxiv, 110, 112
Battery Mounting System, 5
Battery Open Circuit Voltage, 90
Battery Room, 129
Belt, 138
Bi - Directional, 114
Bill Meter, 99
BIS, xvii, 40-41
Blocking Diode, 39, 61, 87-88
BOOT, xvii, 19, 43
Building, v-x, 2, 4, 11, 131
Bulb, 96-97
Bypass Diode, 61, 87, 116

C

C-Rating, xxiii, 127-128
Cable Gland, 125
Cable Loss, xxv, 30, 41, 115
Cables, 43, 65, 123-125
Cadmium Telluride, xvii, 25
Capacity, i-iv, xviii, xxiii, 20, 30, 62, 68-69, 73-76, 78-79, 84, 89-91, 98, 106-108, 111, 117, 127-128
Capacity Test, 117
CAPEX, xvii, 18-19
CEIG, xviii, 28
Cell Area, 6, 54
Central Financial Assistance, iv, xvii, 20
Central Inverter, 38, 92
CFL, xvii, 97
Chalk Line, 47
Charge Controller, xvii, 36, 42, 44
Circuit Breaker, xvii, xx, 43
Civil Layout, 117
Clamp Meter, 109
Clamp Test, 117
Clean Room, 132
Climate Variation, 7
Clip Test, 117
Cloves, 138
Coal, i, iii-iv, 53, 58
Combiner Box, xviii, 39-41
Compass, 47
Concentrator, 34
Concrete Bases, 124

Connected Load, 3, 30, 63
Connection Resistances, 129
Continuity Test, 117
Conversion Efficiency, 7
Copper, xvii, 25, 44, 109, 130
Copper Sulphate, 109
Corrosion, 124, 129
Country, ii-iii, 20-21, 149
Crack, 129
Crimper, 47
Crystalline Silicon, 25-26
Current, xvii-xix, xxiii-xxv, 5-6, 22-23, 42, 49-50, 52, 54-57, 59-61, 69, 71-72, 77, 79-81, 83-88, 90-91, 94-96, 98-107, 109-110, 112-113, 116, 119, 127, 129, 131
Current Meter, 99
Cut In - Out Voltage, 89
Cut Off Voltage, 89
Cut On Voltage, 89
Cutters, 47
Cutting, 131

D

Daily Solar Radiation, 8
Danger Sign, 139
Data Acquisition Systems, xviii, 39
Day Light Falling Angle, 6
DC Circuit, 23, 95
DC Combiner Box, 41
Degrade, 35
Diagnosis of Malfunctions, 121

Diameter, xix, 49-50, 67-68
Diffuse Solar Radiation, 2
Digital Ammeter, 98
Digital Camera, 47
Direct Solar Radiation, 1
Direction of Sun, 4
Disadvantage, 22
Discharging, 72, 74, 88-89, 91, 108
Discharging Time, 72, 74
Distribution Board, 44, 65
Distribution Transformer, xviii, 30, 98
DOD, xviii, xxiii, 62, 78, 88-89, 105, 107, 111, 126
Double Axis Mechanical Tracker, 4
Drill Machine, 47
Dust, 30, 137
Dust Storm, 137

E

Ear Plug, 138
Earth, 1-2, 42, 44, 123, 125, 129-131
Earth Pit, 42, 123, 125
Earthing, 43, 123, 125, 129-132
EB Substation, 2
Electric Current, 49
Electrical Appliance, 96
Electrical Circuit, 49-50, 98-99
Electrical Energy, xxiii, 33, 37-38, 49, 51, 88
Electrical Inspector, xviii, 27
Electrical Layout, 117

Electrical Power, 49
Electrical Voltage, 49
Electricity Bill, xviii, 3, 120
Electrolyte, 48, 125
Electrons, 49
Employee, 138
Energy Meter, 99, 113-115
Ensure Reliability, 66
Entrepreneur, viii-ix, xv, 15-18, 154
Equipment Safety, 131
ESCO, xviii, 19
Explosive Symbol, 139
Exported Energy, 29
Extension Boards, 47

F

Factory, 79, 139
Failure Occur, 124
Fan, 65, 77-78, 109-111, 120
Farad, 50, 94
Fill Factor, xviii, xxiii, 26, 58, 60, 86-87
Fire Station, 139
First Aid Box, 139
Fixed Structure, 2, 4
Flammable Materials, 125
Force, 49
Formation of Hot Spots, 87
Forward, 87, 113-114
Forward Direction, 113
Frequency Fluctuation, 39
Frequency Variation, 121
Fridge, 109

Fuse Puller, 47
Future Load, 97

G

Gain Certainty, 66
Gallium Indium Phosphide, xviii, 25-26
Galvanization, 36
Gauze Pads, 139
Glass, 130
Global Solar Radiation, 1
Goggle, 48, 138, 140
Greenhouse Gases, iv, 22
Grid - Connected Solar, 23-24
Grid Connection, 121
Grinder, 109
Gross Meter, 29
Ground, x, 42, 44, 128, 130
Ground Fault, 42

H

Hack Saw, 47
Hazardous Area, 139
Heat, 6, 33, 54, 58, 60, 63, 124
Heat Dissipation, 63
Heat Energy, 6, 33
Heat Falling, 6
Heavy Rain Falls, 137
Helmet, 138-140
Henry, 50, 94
Higher Temperature, 127
Hole Saw & Punch, 47
Hospital, 139

Hot Spot, 6, 87
Human Safety, 84, 131
Hybrid Solar PV System, 23-24
Hydro Plant, 54
Hydrometer, 48, 117
Hydrometer Test, 117

I

IEC, xix, 34-35, 39-41, 152
Imported Energy, 29
Industry, v-xi, 138-139, 150
Inspection, xi, 65-66, 71, 118
Installed Capacity, i, iv, 20
Insulation Resistance, 119, 130
Internal Rate of Return, xix, 135
Inverter, xxiii-xxv, 5, 24, 27, 30, 34-42, 44, 62, 65-66, 90, 92-94, 109, 111, 115-117, 119, 121, 124-125, 128-130, 137
Inverter Capacity, 111
Inverter Efficiency, xxiii, 93, 111, 119
Inverter Mounting System, 5
Inverter Open Circuit Voltage, 90
Inverter Synchronization, 121
IP, xviii, 40
IREDA, xix, 14, 152
Iron, xviii, 33, 130
ISO, xviii, 40

J

JNNSM, iii, xix, 13
Joule, 51

K

KW, xix, 1, 8-9, 51-53, 68-69, 74, 79-81, 92, 96-97, 111-112, 115

KWh, xix, 8, 30, 33, 51, 54, 73-74, 96-97, 107, 110, 115, 120

L

Latitude, 2-4

Lead Acid, 37, 72-76, 79, 112, 127

Lead Acid Battery, 72-76, 79, 112, 127

Leak, 129

Life Cycle Assessment, xix, 133-134

Life Cycle Cost, xix, 22, 133-134

Life of Battery, 90, 125-126

Light Falls, 6-7

Light Intensity, 6

Lighting Arrestor, xix, 41-42, 44, 131

Lightning Protection, 123

Line Dori, 48

Liquid Vented Lead Acid, 127

Lithium Ion, 37

Lithium Ion Polymer, 37

Load, xxv, 3, 5, 25, 30, 36, 52, 62-63, 77, 89, 96-97, 109, 111, 113-115, 124

Location, 3-5, 63, 66

Longitude, 3

LT Panel, 139

M

Maintenance Free, xxi, 37, 128

Maintenance Manuals, 81

Maintenance of PV, 123

Manager, 138-139

Manuals, 45, 81, 130

Maximum Open Circuit Voltage, 6

Maximum Power, xviii-xx, xxii-xxiii, 7, 11, 27, 36, 51, 58-60, 85, 99, 119

MC4, xix, 36, 41, 65

MCB, xx, 43

Measuring Tape, 47-48

Mechanical Damage, 90

Mechanical Layout, 117

MEP, ix-xi, xix, 18

Meter, xxi-xxii, 24, 28-29, 39, 62, 94, 99, 109, 113-115, 123, 128

Micro Inverter, 38, 92

Minimum Power, 7, 27, 60

MNRE, iii-iv, xx, 13, 29

Module Soiling Loss, 8

Module Surface Temperature, 116

Module Temperature Loss, 8

Momentum, 49

Monitoring Equipment, 137

Monitoring System, 117, 125

Month, 9, 27, 127

Motor, 96, 101

Mounting Structure, 4, 124, 128

MPPT, xix, 24, 27, 36, 38, 42, 44

MS Excel, xix
MS Word, xix
Multimeter, 47, 98-100, 109, 116, 119
MW, iv, x, xx, 8-9, 33, 52-53, 92
MWh, xx

N

NAPCC, xx, 13
National Solar Mission, iii, xix, 13, 20
Nearest Load Center, 3
Negative, 99, 101, 113-114
Net Meter, 24, 29, 39, 62, 114-115
Net Metering Regulation, 30
Neutrons, 49-50
Nickel Cadmium, 37
Nickel Metal Hydride, 37
No Galvanisation, 124
No Shadow, 87
No Sunlight Resistant, 124
Non Shaded Cell, 5-6
Northern hemisphere, 1
Nose Mask, 138
NSDC, xx, 14, 28
NSQF, xv, xx, 14-17

O

OCB, xx, 43
Occupational Health and Safety, xx, 138

Ohm, xix, 50, 94, 125, 128-130
Oil and Gas, 53
Old Technology, 128
One Time Battery, 88
Open Circuit, xxii, 6, 57, 59-61, 66, 71, 83-87, 90, 101-102, 104-105, 108, 116
Operating Manuals, 45, 81
Operating Temperature, 6-7, 66-67
OPEX, xx, 18-19
Opposite Polarity Terminals, 101
Optimum Output, 1
Over Current, 42
Overcharging, 88

P

Parallel, xxiv-xxv, 22-23, 59, 71-72, 74-76, 80-81, 83, 86-88, 91, 98-102, 104, 112-113, 116, 120
Parallel - Series, 74-75, 86, 120
Pay Back Period, 135
Peak, xxii-xxiii, 9-11, 25, 53-54, 58, 98
Peak Output Power, 9-11
Performance Monitoring, 121
Performance Warranty, 35
Personal Protective Equipment, xx, 137
Petroleum, 53
Phase Unbalance, 71, 121
Physical Damage, 125

Pipe, 131
Plate, 35, 131
Pliers, 47
Plumb Bob, 48
Polarity, 87, 101, 116-117, 119
Positive, 101, 113-115
Positive Terminal, 101
Power Consumption, 96
Power Factor, xxiii, 95
Power Loss, xxiv-xxv, 39, 79, 115
Power Sector, 20-21
PPA, xx, 30
Prevent Leakage Current, 87
Prevent Reverse Current Flow, 87
Price of Components, 5
Project Report, 81
Proper Monitoring System, 125
Protons, 49-50
Pump, 109
PV Array, xxiv-xxv, 58-59, 65, 79-81, 115, 125, 128
PV Mounting Structure, 4
PV Output, 4-5, 7, 36, 120
PV Plant, xx, 2-5, 28-29, 42-44, 53-54, 63, 66, 81, 119, 121, 123-126
PV String, xxiv, 40, 80-81, 83, 113
PV System, v-vii, 1, 3-5, 7, 23-25, 33, 36-37, 41-45, 47, 49, 60-61, 83-84, 88, 117-118, 123, 130, 135
Pyranometer, 33, 41, 119

Q

Qualification Pack, v-xi, 14-18
Quality Assurance, xx, 118
Quality Control, xx, 118
Quality of the PV System, 5

R

Real Power, xxiii, 95
Rectifier, 36-38, 90, 109
Rectifier Open Circuit Voltage, 90
Redo, 118
Reflected, 1-2
Reflective Jacket, 140
Renewable Energy, iii-iv, xi, xix-xxi, 13-14, 19-21, 53, 149-156
Repair, 118, 137
Replace, iii, 118
Resistance, xxi, xxiv, 49-50, 67-68, 94-95, 98, 100, 110, 119, 128-130
Resistivity, 44, 94-95
Return on Investment, xxi, 135
Reverse, 61-62, 87-88, 113-114
Reverse Direction, 113
Reverse Flow of Current, 88
Reverse Power, 61-62, 88
Reverse Voltage, 88
RMS, xxi, 94-95, 98
Rod, 131
Roof Perforation, 124
Roof Pitch, 2-4
Roof Top Installation, 28

Rooftop Solar Grid Engineer, x-xi, xv, 15-18, 154
Rooftop Solar Photovoltaic, viii-ix, xv, 15-18, 154
Routine Cleaning, 125
Rubber, 48, 130
Rubber Apron, 48
Rubber Gloves, 48
Rust, 123

S

Safety Cloth, 138
Safety Goggles, 48, 140
Safety Harness, 140
Safety Manuals, 130
Salt Water, 124
Same Polarity Terminals, 101
Scattered, 2
SCGJ, iv, vi-xi, xv, xxi, 14-15, 28
Scheduled Maintenance, 125
Scratch, 65
Screwdrivers, 47
Sea Level, 2
Sealed Lead Acid Batteries, 127
Sealed Lead Acid Battery, 72
Sealed Maintenance Free, xxi, 37
Seasonal Tilt, 27
SECI, xxi, 13, 154
Selenide, xvii, 25
Self Consumption, 29
Sensor, 34, 123
Series, xxiv-xxv, 22-23, 59, 71-72, 74-76, 80-81, 83-88, 91, 98-103, 105, 112, 120

Set In Order, 140
SGJ/Q0101, v, xv, 15-17
SGJ/Q0102, vi, xv, 15-17
SGJ/Q0103, vii, xv, 15-17
SGJ/Q0104, viii, xv, 15-16
SGJ/Q0105, ix, xv, 17
SGJ/Q0106, xi, xv, 16-17
Shaded Cell, 5-6
Shading Obstacles, 4
Shadow, 2-4, 30, 87
Shadow Analysis, 3-4
Shine, 140
Shoe, 138-139
Short Circuit, xviii, 42, 56-57, 59-61, 71, 83-87, 102, 104-105, 116-117, 119
Single Axis Mechanical Tracker, 2
Siren Alarm, 139
Site Feasibility Study, 3
Site Layout, 4
Site Survey, v-vii, 1-2, 11, 30, 81
Size of the Conductor, 129
Small Flashlight, 48
Snow Fall, 30
SoC, xxi, xxiii, 89-90, 106-107, 116
Sodium Sulphate, 109
Soil Deposit, 7, 30, 124
Soil Quality, 2-3
Soil Test, 3
Solar Cell, 5-7, 24-26, 55, 57-59, 66-67, 84-85, 87, 100
Solar Collector, 33
Solar constant, 1

Solar Panel, 1, 26, 33, 35, 59-60, 67-68, 133-134
Solar Proposal Evaluation, ix-x, xv, 15-18, 155
Solar PV Installer - Civil, vii, 14-16, 18, 155
Solar PV Installer - Electrical, vi, 14-16, 18, 155
Solar PV Installer - Suryamitra, 14-16, 18, 155
Solar Radiation, 1-2, 6, 8-9, 24, 30, 33, 58, 119
Solar Trackers, 27
Solar Water Heater, 33
Sort, 140
Southern hemisphere, 1
Spanners, 47
Specialist, ix-x, xv, 15-18, 155
Specific Gravity, 109, 116, 129
SSC, vi-viii, xxi, 18
Standalone Solar PV System, 23-24
Standard Operating Procedure, xxi, 140
Standardize, 140
State of Charge, xxi, 89
STC, xxi, xxiii, 7, 9-11, 66-67, 99-100, 108
String, xxiv, 38, 40, 59, 65, 68-69, 71, 80-81, 83, 92, 112-113
String Inverter, 38, 40, 92
Strong Wind Situation, 137
Sulphur Hexafluoride, xxi, 43

Sulphur Oxide, 109
Sulphur Phosphate, 109
Supervisor, 138
Surge Protection Device, xxi, 42-43
Sustain, ii, 140
Switchgear, 42

T

Tariff Structure, 3
Technical Manuals, 130
Test and Commission, v-vi, 83
Thermal Plant, 54
Thermocouple, 116
Thermometer, 34, 43
Thin Film Solar Cell Efficiency, 25
Tilt Angle, 2-4, 7
Torque Wrench, 47
Trainer, vi-xi, 18
Transformer, xviii, 27, 30, 40, 43, 62, 66, 98
Transmission Losses, 30
TV, 109
Type of Roof, 3

U

UL, xxi, 35
Under Voltage, 42
Uni - Directional, 114
Unlimited Source, 21
UV Resistant, 124

V

V-I Characteristics, 22
V-P Characteristics, xxi, 22
Vegetation, 65
Ventilated Room, 131-132
Ventilation, 124-125
Vernier Caliper, 48
Voltage, xviii-xix, xxi-xxv, 5-6, 22, 30, 42-43, 49-50, 52, 55, 57, 59-63, 66-69, 71-81, 83-92, 94-95, 98-113, 116-117, 119, 121, 129-130
Voltage Drop, xxii-xxiii, 30, 50, 67-68, 95, 109-110
Voltage Fluctuation, 121, 130
Voltmeter, 98-99, 109

W

Washing Machine, 109
Water Supply Connection, 2
Watt - Hour Meter, 99
Watt Second, 51
Wattmeter, 98-99, 119
Weather Condition, 5
Weather Dependent, 22
Welding, 131, 138
Wind Direction, 34
Wind speed, 2-3, 33-34, 36, 124
Wire Colour, 44
Wire Strippers, 47
Wood, 130
World Leader, 21

www.ingramcontent.com/pod-product-compliance
Lightning Source LLC
Chambersburg PA
CBHW030938180526
45163CB00002B/613